Foundations of Heat Transfer

Foundations of Heat Transfer

Editor

Dorin Vasilescu

Foundations of Heat Transfer

Edited by **Dorin Vasilescu**

Printed in 2017

ISBN: 978-1-68117-120-3
Library of Congress Control Number: 2015951143

© 2016 by
SCITUS Academics LLC,
616, Corporate Way, Suite 2, 4766,
Valley Cottage, NY 10989

www.scitusacademics.com

Preface

Heat may be defined as the transfer of thermal energy across a well-defined boundary around a thermodynamic system. The thermodynamic free energy is the amount of work that a thermodynamic system can perform. Enthalpy is a thermodynamic potential that is the sum of the internal energy of the system plus the product of pressure and volume. Heat transfer is the exchange of thermal energy between physical systems, depending on the temperature and pressure, by dissipating heat. The fundamental modes of heat transfer are conduction or diffusion, convection and radiation. Heat transfer always occurs from a region of high temperature to another region of lower temperature. In the simplest of terms, the discipline of heat transfer is concerned with only two things: temperature, and the flow of heat. Temperature represents the amount of thermal energy available, whereas heat flow represents the movement of thermal energy from place to place.

Heat transfer is a process function, as opposed to functions of state; therefore, the amount of heat transferred in a thermodynamic process that changes the state of a system depends on how that process occurs, not only the net difference between the initial and final states of the process. Several material properties serve to modulate the heat transferred between two regions at differing temperatures. Examples include thermal conductivities, specific heats, material densities, fluid velocities, fluid viscosities, surface emissivities, and more. Taken together, these properties serve to make the solution of many heat transfer problems an involved process.

Foundations of Heat Transfer focuses on the basic modes of heat transfer.

Table of Contents

CHAPTER 1

Effect of Spacecraft Aerodynamics and Heat Shield Characteristics on Optimal Aeroassisted Transfer

Antonio Mazzaracchio, Mario Marchetti

Astronautical, Electrical, and Energetic Engineering Department, Sapienza University of Rome, Rome, Italy

Keywords

Aeroassisted Maneuver; Heat Shield; Optimization; Orbital Transfer; Thermal Protection System

ABSTRACT

A spacecraft designed to operate in a planetary atmosphere must have an adequate heat shield to withstand the high heat fluxes and heat loads that are generated by aerodynamic heating. Very often, the mass of the thermal protection system is a significant fraction of the total mass of the vehicle. In contrast, performing maneuvers in the atmosphere, that would be very costly in terms of propellant consumption if they were performed completely outside of the atmosphere in a classic way, is a very attractive prospective technique. The advantages and disadvantages in terms of total mass spared must be determined. The mission investigated involves an aeroassisted coplanar transfer from a high to a low Earth orbit. The approach uses a combination of three propulsive impulses in space together with an aerodynamic maneuver in the atmosphere. The heat shield adopted is fully ablative, given the expected high values of the entering heat flux. The convenience of the aeroassisted maneuver and the influence of the parameters involved are evaluated in comparison to a conventional Hohmann transfer. In particular, a parametric analysis is performed by varying the following characteristics of the vehicle: aerodynamic efficiency, mass-to-surface ratio, deorbit impulse, and initial altitude of the orbit. The influence of the thermal protection system is examined by assessing the impact of the type of ablative material employed, the thermal safety factor, and the allowable temperature for the adhesive layer on the substructure. The analysis is conducted with a highly representative thermal model by coupling the dynamic and thermal analyses and using a genetic optimizer. The optimization methodology and the thermal model are completely original. The results indicate the importance of choosing low-density ablative materials, of adopting a suitable thermal safety factor, and of choosing high-performance adhesives. The optimal trajectories obtained correspond to a zero second propulsive impulse.

INTRODUCTION

One of the future objectives in space activities is to use aeroassisted orbital maneuvers, i.e., maneuvers that are carried out with the aid of the atmosphere, to satisfy the increasingly stringent constraints of the cost of space missions. Depending on the goal of the mission, aeroassisted maneuvers are performed in different ways: aerogravity assist, aerobraking, and aerocapture. To these maneuvers, one must add the re-entry maneuvers and some possible variants such as the skip re-entry technique. These maneuvers are very attractive, although they are only theoretical at this point (except for rare uses in aerobraking), because they can substantially reduce the requirements of a space mission in terms of propulsion and flight time in favor of, among other things, the possibility of housing a larger payload.

Underlying this approach is the possibility of exploiting the presence of the atmosphere of the celestial body around which one wants to operate to lower the overall energy required. In practice, one tries to execute the complete operation with the help of aerodynamics because orbital maneuvers are quite expensive in terms of propulsion, especially those outside the orbital plan. Thus, the design of a spacecraft with specific reference to atmospheric portion of flight must present an efficient aerodynamic configuration; however, the priority constraint of the overall cost of space missions must be satisfied.

From this brief introduction, it is already clear that once shown the technical feasibility of the aeroassisted maneuver, one must evaluate its convenience compared with alternative hypotheses, such as a classical purely propulsive maneuver. In fact, optimizing a spacecraft and its mission, or more specifically its trajectory for a given mission, is always a compromise between the interests of performance, security, and economics, which are almost always in mutual conflict.

Usually, minimizing the mass of the heat shield while respecting the limits of safety is a primary requirement. Indeed, any savings in terms of the mass of the thermal protection system (TPS) can be translated into an increase of the "useful" mass, namely, into an opportunity to accommodate a greater payload or extend other spacecraft subsystems. The problem becomes more complex when considering an aeroassisted maneuver and taking into account the mass of propellant still required during some of the various phases of the maneuver itself.

In this regard, a basic scheme is usually adopted in the literature to perform an aeroassisted maneuver—the use of propulsion only in space together with one or more atmospheric segments of pure aerodynamic flight—which is also considered in this work. Maximizing the benefits at the propulsive level for an aeroassisted maneuver normally requires a more intensive use of the atmospheric phase of flight. Thus, the resulting trajectory touches the denser layers of the atmosphere with longer crossing times. Therefore, a larger TPS with its relative higher mass fraction must be adopted.

A legitimate question that arises at this point is whether the resulting increase in the mass of the TPS, together with the mass of propellant required to enter the atmosphere and then to achieve the final orbit, may override the convenience of the aeroassisted maneuver compared with a classical operation, which is a purely propulsive extra-atmospheric maneuver. It seems evident that this is an optimization problem.

All of the above considerations concerning both the evaluation of the convenience of the aeroassisted approach and the search for the optimal solution and the methodological sphere itself are the basis of the research questions that motivated this work.

An original procedure was carried out jointly with the implementation of software developed by the authors to optimize the aeroassisted orbital maneuvers using a genetic algorithm (GA) with simultaneous evaluation of the optimal configuration of the associated heat shields and the coupling of the dynamic and thermal analyses. The tool was verified by comparing its results with those found in the literature. This tool, because of its level of implementation details, is suitable for the conceptual development stages of a spacecraft and its mission.

The initial intention was to evaluate the influence of various parameters — orbital, aerodynamic, and dimensional — on the feasibility and convenience of the mission. For an introductory analysis of the problem and to evaluate the importance of various factors, coplanar transfer (aerobraking) from a high Earth orbit (HEO) to a low Earth orbit (LEO), both circular, is proposed as a case study. The vehicle is a delta wing spacecraft that is protected by an ablative TPS with uniform thickness. The problem is studied in the absence of constraints on the maximum allowable entering heat flux.

Beyond the presentation of the problem in Section 1, Section 2 describes the model and the optimization procedure. The case study is presented in Section 3, and the relevant results and analyses are discussed in Section 4. Finally, Section 5 offers a summary, conclusions, and recommendations for future improvements.

MODELS AND OPTIMIZATION

The description of the models, the governing equations, and the relevant assumptions for the problem—i.e., thermal models, aerodynamic heating, atmospheric flight mechanics, and heat shield configuration—are thoroughly presented in [1] and [2]. Reference [2] can be consulted for full details of the original optimization procedure.

All of the analysis presented here was performed using the cited software developed by the authors called ATHSHO (Aeroassisted Trajectory and Heat SHield Optimization). It is important to recall that thermal analysis is performed with a one-dimensional plane model and that the adopted GA refers to a mixed one-point/two-point crossover operator together with a reproduction plan that provides a full generational replacement with elitism [3-5].

CASE STUDY

A parametric analysis was performed by varying the following characteristics of the vehicle: aerodynamic efficiency, mass-to-surface ratio, deorbit impulse, and the altitude of the initial orbit. The influence of the characteristics of the TPS was examined by assessing the type of ablative material adopted, the thermal safety factor, and the allowable temperature for the bond-line, i.e., the adhesive layer on the substructure.

Hypotheses

The main hypotheses considered are the following:

- The initial total mass of the vehicle is given.
- The attitude control is accomplished only through the angle of attack.
- The entering heat flux is unconstrained.
- The TPS is fully ablative with uniform thickness.

Vehicle

The vehicle model is a delta wing shuttle with a high L/D ratio, which is comparable in the first instance to the configuration and dimensions of the Boeing X-37A vehicle (**Figure 1**).

The dimensions, sizes, and aerodynamic characteristics of the vehicle used were taken in part from [6-8]. Other data were reasonable assumptions made by the authors. The main dimensions and characteristics of the vehicle are listed in Table 1, and the principal aerodynamic and propulsive parameters are listed in Table 2.

The values shown are those considered for the nominal reference case. Some of them may vary depending on the purpose of the study (Section 4).

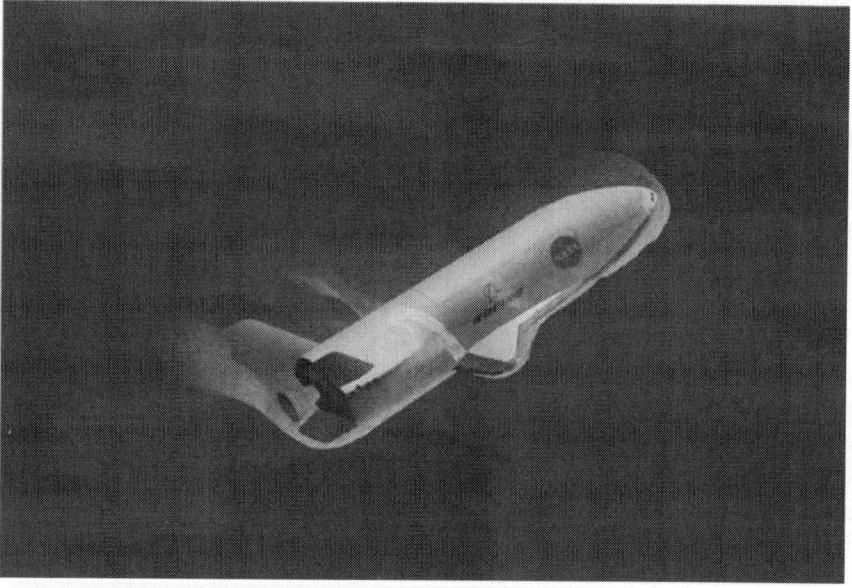

Figure 1: Boeing X-37A (source NASA.gov).

Table 1: Vehicle dimensions and characteristics

Vehicle length	l_{ve}	9.38 m
Vehicle body radius	r_b	1.00 m
Vehicle wing span	ws_{ve}	4.50 m
Vehicle wing cord	wc_{ve}	3.50 m
Vehicle reference surface	S	11.69 m^2
Vehicle TPS total surface	$S_{TPS,ve}$	42.65 m^2
Bond-line limit temperature	$T_{BL,lim}$	450 K
Thermal safety factor	TSF	1

Table 2: Vehicle's aerodynamic and propulsive characteristics

Zero-lift drag coefficient	C_{D0}	0.032
Induced drag factor	K_D	1.4
Lift coefficient derivative	$C_{L,a}$	0.5699
Maximum lift coefficient	$C_{L,max}$	0.4
Propellant specific impulse	I_{sp}	310 s

Mission and Maneuver

Table 3 lists the values of the altitudes for the initial Geostationary Earth Orbit (GEO) and final LEO for the aerobraking maneuver as well as the conventional height assumed for the atmosphere.

Even in this case, the indicated values are those used for the reference case. In particular, in the parametric analysis performed, the value of the initial HEO altitude was varied. The physical properties of the atmosphere were derived from the model 1976 US Standard Atmosphere [9].

Figure 2 shows a classic schematic for a HEO-LEO aerobraking maneuver. The strategy involves the combined use of aerodynamic maneuvering in the atmosphere and some extra-atmospheric propulsion phases. More precisely, one assumes that the propulsive phases are concentrated in three impulses in space and that the portion of atmospheric flight is performed without the

Table 3: Maneuver characteristics

Initial HEO (GEO) altitude	H_A	35,786 km
Final LEO altitude	H_B	480 km
Atmosphere's upper limit	H_{atm}	129.6 km

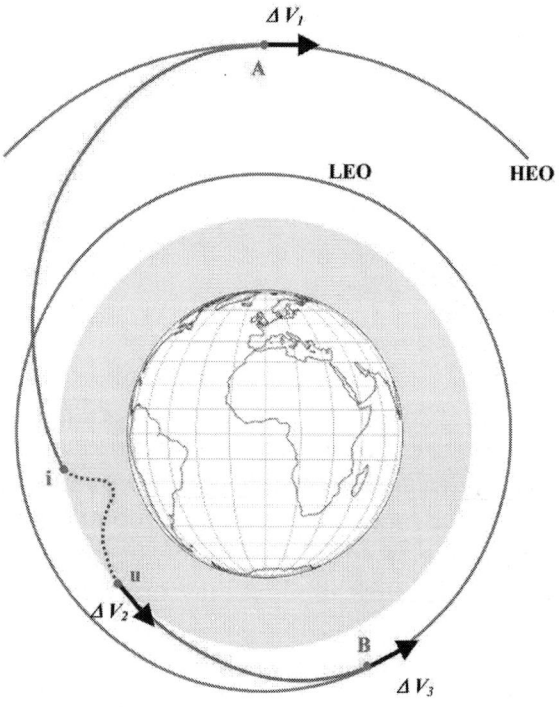

Figure 2: Schematic of a HEO-LEO aerobraking maneuver.

- Existing adhesives with enhanced characteristics allow the threshold temperature to be increased to 400°C, which results in greater heat shield thickness savings [10].
- Ablative material: PICA-15 (hereinafter referred to as PICA), AVCOAT 5026-HCG (hereinafter referred to as AVCOAT), and FM 5055 CP in its Reduced Density version (hereinafter referred to as RDCP).

Among all the possible combinations of these parameters with their associated values, 12 cases were analyzed for each value of the M/S ratio for a total of 36 different scenarios. The combinations analyzed are presented in Tables 4-6, where the index (a, b, and c) marks the adopted value for the load. Moreover, because the range of variation of ΔV_1 is analyzed by steps of 10 m/s, 390 different cases form the database of results for the analysis of the case study in question.

As a reference, the nominal case is chosen for the GEOHEO transfer that provides a TPS made of PICA, with $T_{BL,\,lim}$ = 450 K, TSF = 1, and ε = 1. The twelve combinations chosen and the corresponding purpose of

Table 4: HEO-LEO transfer, parameters and values (cases 1, 2, 3, and 4)

		Case 1a	Case 1b	Case 1c	Case 2a	Case 2b	Case 2c	Case 3a	Case 3b	Case 3c	Case 4a	Case 4b	Case 4c
HEO-LEO													
HEO = GEO	35,786 km - 480 km	✷	✷	✷	✷	✷	✷	✷	✷	✷	✷	✷	✷
M/S													
200 kg/m²	$M_{sx,ref}$ = 2338.0 kg	✷			✷			✷			✷		
300 kg/m²	$M_{sx,ref}$ = 3507.0 kg		✷			✷			✷			✷	
450 kg/m²	$M_{sx,ref}$ = 5260.5 kg			✷			✷			✷			✷
Aerodynamic efficiency													
Medium 1.5	C_{D0} = 0.1; K = 1.111	✷	✷	✷	✷	✷	✷				✷	✷	✷
High 3.0	C_{D0} = 0.017; K = 1.76							✷	✷	✷			
ΔV_1													
	1490 - 1590 m/s	✷											
	1490 - 1560 m/s		✷										
	1490 - 1520 m/s			✷									
	1490 - 1590 m/s				✷								
	1490 - 1550 m/s					✷							
	1490 - 1520 m/s						✷						
	1490 - 1710 m/s							✷					
	1490 - 1600 m/s								✷				
	1490 - 1560 m/s									✷			
	1500 - 1600 m/s										✷		
	1490 - 1560 m/s											✷	
	1490 - 1520 m/s												✷
Ablative material													
	PICA-15	✷	✷	✷	✷	✷	✷	✷	✷	✷	✷	✷	✷
T-bond line													
	450.00 K	✷	✷	✷				✷	✷	✷	✷	✷	✷
	673.15 K				✷	✷	✷						
Thermal safety factor													
	1	✷	✷	✷	✷	✷	✷	✷	✷	✷			
	2										✷	✷	✷

Table 5: HEO-LEO transfer, parameters and values (cases 5, 6, 7, and 8).

		Case 5a	Case 5b	Case 5c	Case 6a	Case 6b	Case 6c	Case 7a	Case 7b	Case 7c	Case 8a	Case 8b	Case 8c
HEO-LEO													
HEO = GEO	35,786 km - 480 km	*	*	*	*	*	*						
HEOH = 1.5 GEO	53,679 km - 480 km							*	*	*	*	*	*
M/S													
200 kg/m²	$M_{ve,bd}$ = 2338.0 kg	*			*			*			*		
300 kg/m²	$M_{ve,bd}$ = 3507.0 kg		*			*			*			*	
450 kg/m²	$M_{ve,bd}$ = 5260.5 kg			*			*			*			*
Aerodynamic efficiency													
Medium 1.5	C_{D0} = 0.1; K = 1.111	*	*	*	*	*	*	*	*	*	*	*	*
ΔV₁													
	No solution	*											
	No solution		*										
	1500 - 1640 m/s			*									
	1490 - 1570 m/s				*								
	1490 - 1520 m/s					*							
	1490 - 1500 m/s						*						
	1440 - 1490 m/s							*					
	1440 - 1470 m/s								*				
	1440 - 1460 m/s									*			
	1440 - 1480 m/s										*		
	1440 - 1460 m/s											*	
	1440 - 1450 m/s												*
Ablative material													
	PICA-15							*	*	*			
	FM 5055 RDCP	*	*	*									
	AVCOAT 5026-H CG				*	*	*				*	*	*
T-bond line													
	450.00 K	*	*	*	*	*	*	*	*	*	*	*	*
Thermal safety factor													
	1	*	*	*	*	*	*	*	*	*	*	*	*

Table 6: HEO-LEO transfer, parameters and values (cases 9, 10, 11, and 12)

		Case 9a	Case 9b	Case 9c	Case 10a	Case 10b	Case 10c	Case 11a	Case 11b	Case 11c	Case 12a	Case 12b	Case 12c
HEO-LEO													
HEOH = 1.5 GEO	53,679 km - 480 km	*	*	*	*	*	*						
HEOL = 0.5 GEO	17,893 km - 480 km							*	*	*	*	*	*
M/S													
200 kg/m²	$M_{ve,bd}$ = 2338.0 kg	*			*			*			*		
300 kg/m²	$M_{ve,bd}$ = 3507.0 kg		*			*			*			*	
450 kg/m²	$M_{ve,bd}$ = 5260.5 kg			*			*			*			*
Aerodynamic efficiency													
Medium 1.5	C_{D0} = 0.1; K = 1.111							*	*	*			
High 3.0	C_{D0} = 0.017; K = 1.76	*	*	*	*	*	*				*	*	*
ΔV₁													
	1440 - 1530 m/s	*											
	1440 - 1490 m/s		*										
	1440 - 1460 m/s			*									
	1440 - 1520 m/s				*								
	1440 - 1470 m/s					*							
	1440 - 1450 m/s						*						
	1420 - 1880 m/s							*					
	1420 - 1640 m/s								*				
	1420 - 1550 m/s									*			
	1420 - 1880 m/s										*		
	1420 - 1810 m/s											*	
	1420 - 1620 m/s												*
Ablative material													
		*	*	*				*	*	*	*	*	*
	AVCOAT 5026-HCG				*	*	*						
T-bond line													
	450.00 K	*	*	*	*	*	*	*	*	*	*	*	*
Thermal safety factor													
	1	*	*	*	*	*	*	*	*	*	*	*	*

the evaluation are summarized in **Table 7**. Figures 4 to 15 report the trends of the mass gain compared with the "all propulsive" case with varied applied ΔV_1 and parameterized as a function of the three chosen values of M/S, which were analyzed for each scenario. The graph on the left of each figure shows the cited mass gain in absolute terms, whereas the graph on the right presents its performance in terms of the percentage of the initial vehicle's mass.

The mass gain is therefore a measure of the convenience of the aeroassisted maneuver compared with the corresponding Hohmann transfer. When reading these graphics, the definitions of the "range of feasibility" and the "range of convenience" are introduced.

The first one is the interval of ΔV_1 in which the aeroassisted maneuver is feasible, whereas the second ΔV_1 interval is the range in which the aeroassisted operation is more convenient than the "all propulsive" one. As mentioned above, having made the run for all values of ΔV_1 leading to a solution, the range of feasibility is directly readable as the x-axis interval of the definition of the curves themselves.

The range of convenience is by definition the interval in which the curve takes positive values. The range of convenience is always within the range of feasibility. **Table 8** shows, for the value M/S = 200 kg/m² (case "a" of Tables 4-6), the values of the three propulsive impulses applied in each scenario, which in each case give the highest convenience for the maneuver (maxima of the "blue" curves in the graphs of Figures 4 to 15).

It is evident that the utmost convenience for the aeroassisted maneuver corresponds to a value of ΔV_2 that is very close to zero, i.e., in the absence of the boost impulse (case 5a has no feasible solutions for the considered value for the load).

This outcome is valid in general for the other values of M/s. Thus, the trajectory optimization process tends to reduce the vehicle's speed while exiting the atmosphere to the "right" value to reach the final LEO with minimal energy expenditure. Saving this fuel for the second impulse is the basis for the achievement of the highest convenience.

Comparing Figures 4 to 15, one can draw some interesting conclusions:

- The increase of M/S in all cases involves a decrease in the range of feasibility; in particular, the maximum possible ΔV_1 diminishes (the right bound of the range).
- The increase of $T_{BL,\,lim}$ (**Figure 5**) from the standard value to the high-performance value due to the reduced thickness of the TPS improves the maximum achievable convenience (by approximately 2%).

Table 7: HEO-LEO transfer, scenarios analyzed

	Combination	Purpose
1	GEO; PICA; ε medium; $T_{BL,lim}$ standard; TSF = 1	Reference nominal case
2	GEO; PICA; ε medium; $T_{BL,lim}$ high performance; TSF = 1	$T_{BL,lim}$ influence
3	GEO; PICA; ε high; $T_{BL,lim}$ standard; TSF = 1	ε influence
4	GEO; PICA; ε medium; $T_{BL,lim}$ standard; TSF = 2	TSF influence
5	GEO; RDCP; ε medium; $T_{BL,lim}$ standard; TSF = 1	Material influence
6	GEO; AVCOAT; ε medium; $T_{BL,lim}$ standard; TSF = 1	Material influence
7	HEOH; PICA; ε medium; $T_{BL,lim}$ standard; TSF = 1	HEO altitude influence; reference case for HEOH
8	HEOH; AVCOAT; ε medium; $T_{BL,lim}$ standard; TSF = 1	Material influence w.r.t. HEOH
9	HEOH; PICA; ε high; $T_{BL,lim}$ standard; TSF = 1	ε influence w.r.t. HEOH
10	HEOH; AVCOAT; ε high; $T_{BL,lim}$ standard; TSF = 1	Material + ε influence w.r.t. HEOH
11	HEOL; PICA; ε medium; $T_{BL,lim}$ standard; TSF = 1	HEO altitude influence; reference case for HEOL
12	HEOL; PICA; ε high; $T_{BL,lim}$ standard; TSF = 1	ε influence w.r.t. HEOL

Legend: in red: GEO nominal case and changes w.r.t. it; in blue: HEOH nominal case and changes w.r.t. it; in green: HEOL nominal case and changes w.r.t. it.

Table 8: Propulsive impulses in the more convenient maneuvers for "a" cases

Case	ΔV_1(m/s)	ΔV_2 (m/s)	ΔV_3 (m/s)
1a	1530.00	1.38	334.75
2a	1530.00	−1.83	332.91
3a	1500.00	−2.21	375.26
4a	1520.00	0.69	526.53
5a	-	-	-
6a	1520.00	−0.95	502.54
7a	1480.00	2.86	366.44
8a	1470.00	−4.00	568.53
9a	1450.00	−3.51	423.53
10a	1450.00	0.67	421.13
11a	1460.00	−3.56	410.11
12a	1440.00	−2.73	267.69

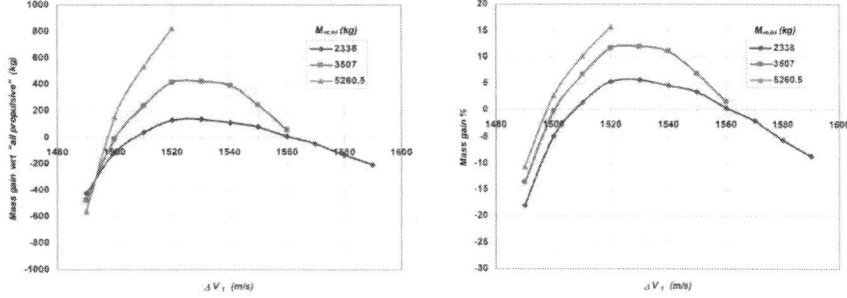

Figure 4: HEO-LEO scenario 1: reference nominal case.

- The increase in the aerodynamic efficiency of the vehicle (**Figure 6**) involves a significant extension of the range of feasibility, but a more moderate increase in the range of convenience. Cases with higher M/S benefit most. Even here, the maximum achievable convenience increases (approximately 2% more).
- The doubling of the thermal safety factor (**Figure 7**), which increases the mass of the TPS, represents a significant performance penalty for the aeroassisted maneuver, which remains affordable for a short interval of ΔV_1 only for the case with highest M/S.
- The use of other ablative materials (Figures 8 and 9),

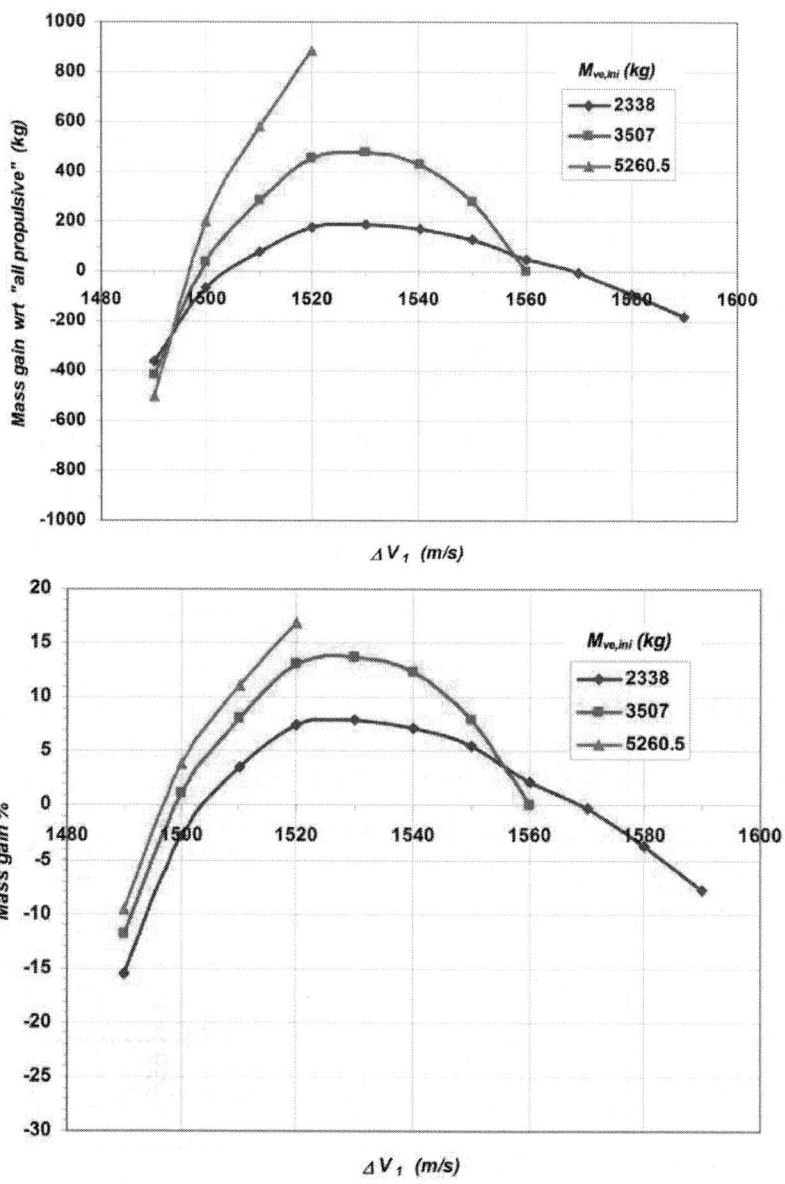

Figure 5: HEO-LEO scenario 2: $T_{BL,lim}$ influence.

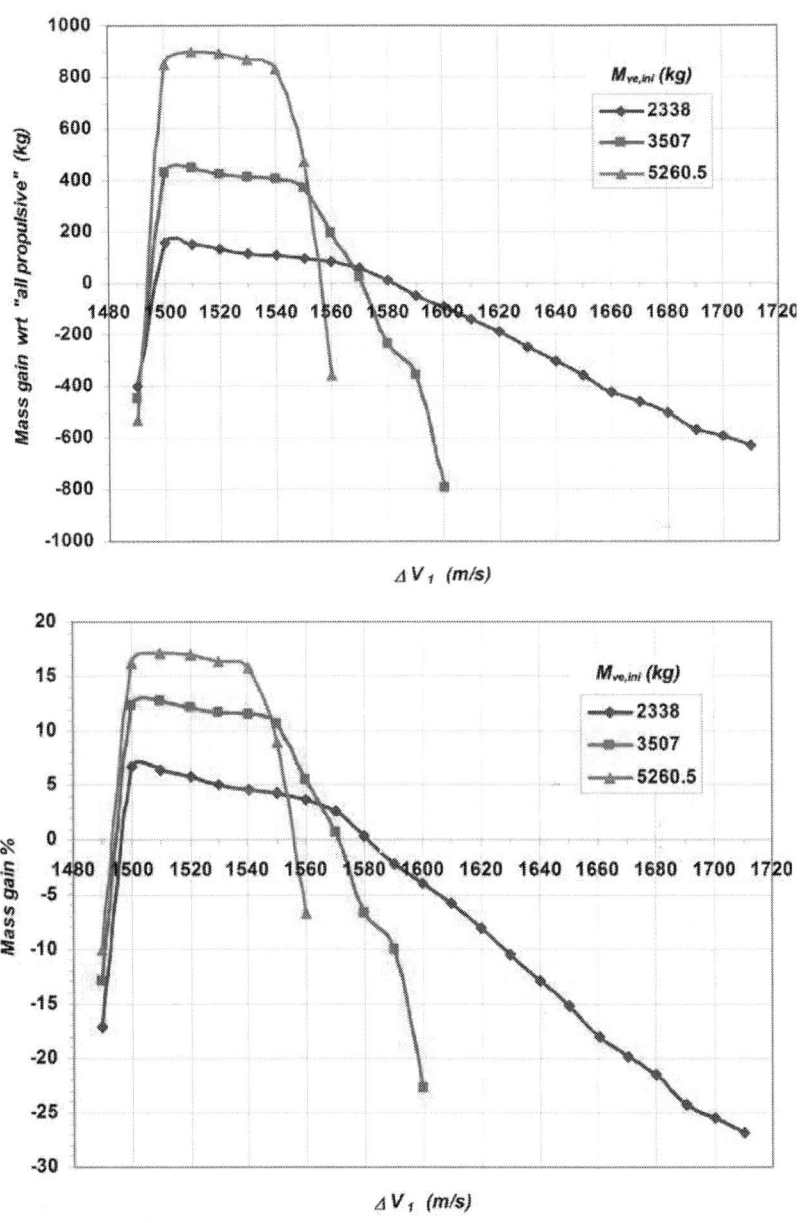

Figure 6: HEO-LEO scenario 3: e influence.

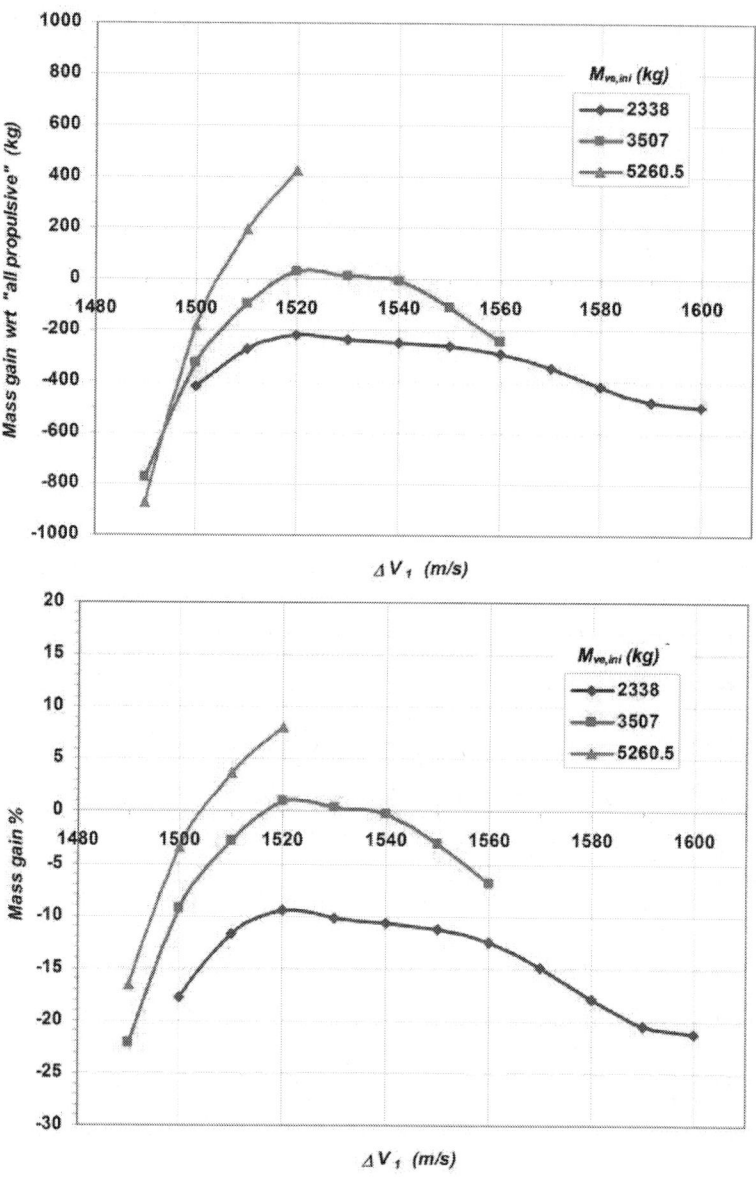

Figure 7: HEO-LEO scenario 4: TSF influence.

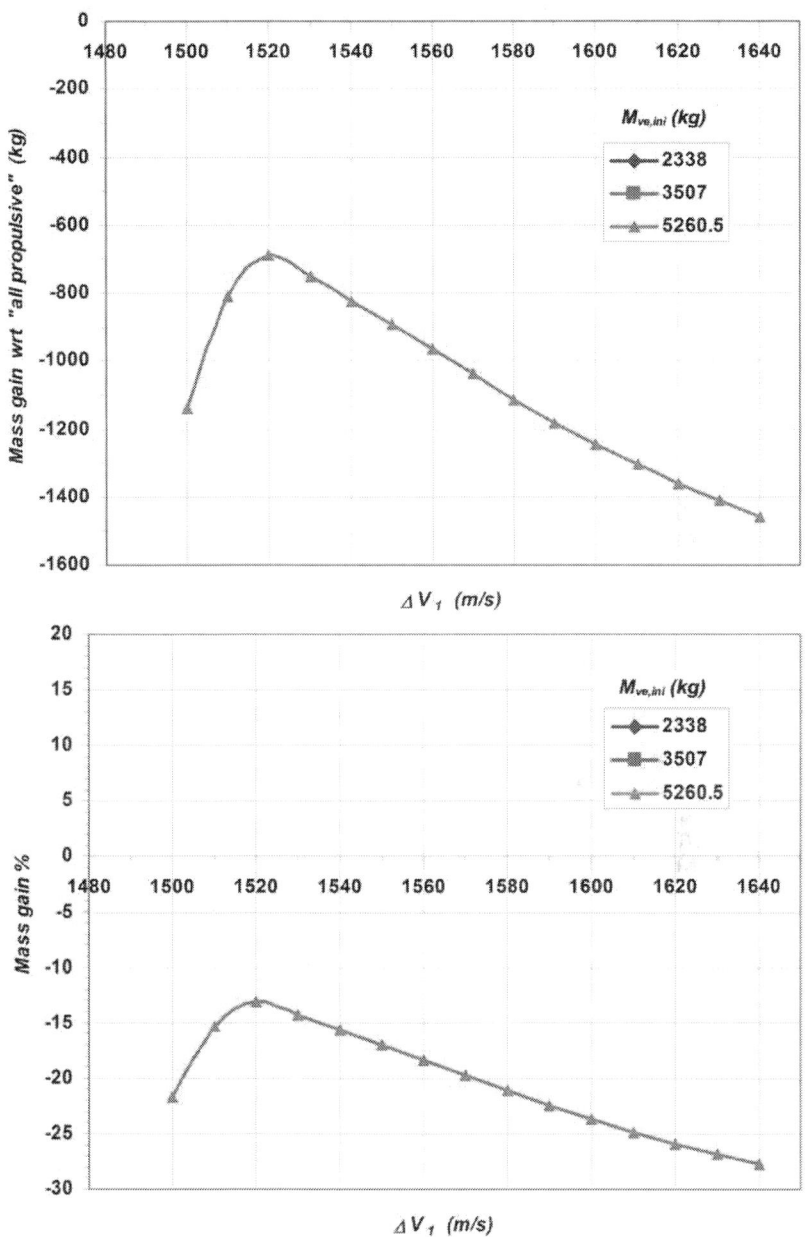

Figure 8: HEO-LEO scenario 5: material (RDCP) influence.

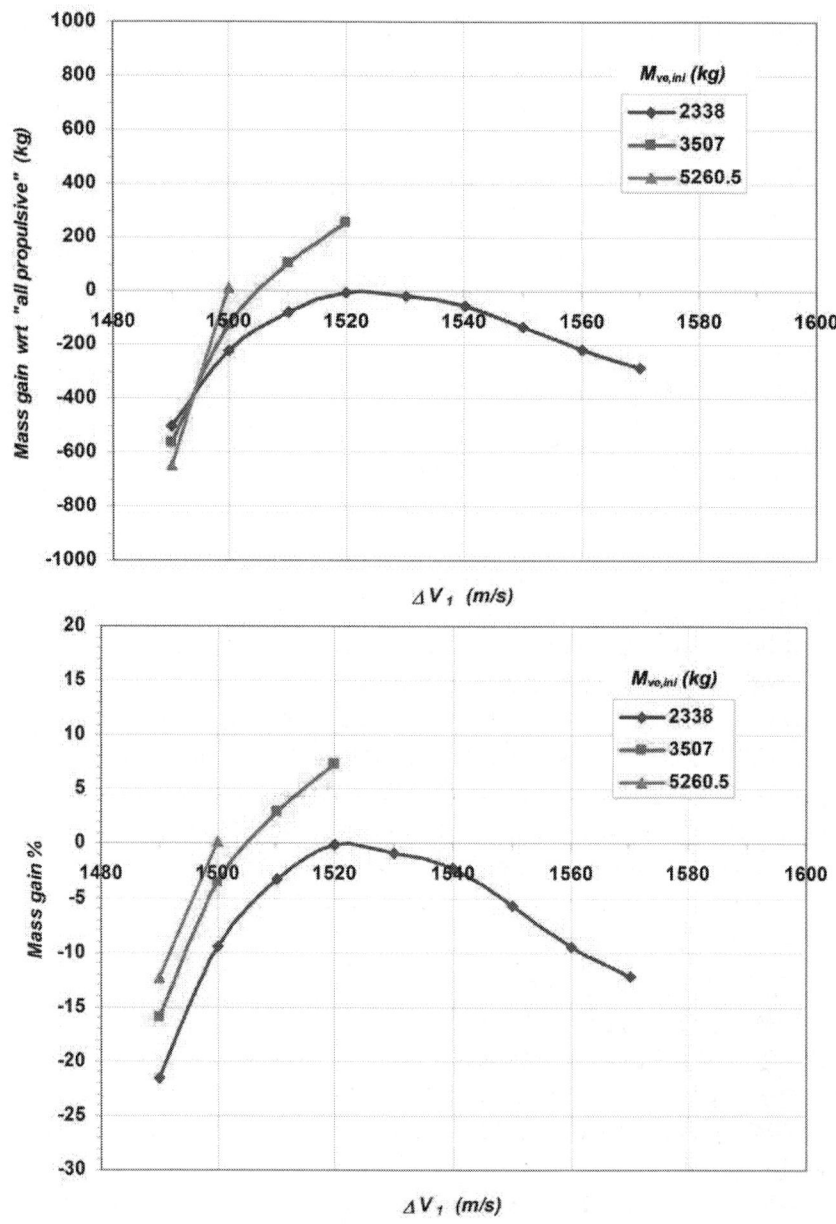

Figure 9: HEO-LEO scenario 6: material (AVCOAT) influence.

compared with the reference material PICA, appears to be inappropriate for a GEO initial altitude. Using the RDCP, the sole case with the highest M/S is feasible, but still not convenient.

Conversely, using the AVCOAT increases the feasibility compared with RDCP but limits convenience compared with PICA. Thus, the relatively low density of the PICA is a discriminating element in the selection of the ablative material. Actually, these results are conservative because of the assumption that the vehicle is covered with a TPS with a uniform initial thickness.

In **Figure 8**, all runs with load values of 200 kg/m^2 and 300 kg/m^2 give results, but the curves are not represented because the final "useful" masses (i.e., all the masses that are not propellant and TPS) are negative. Thus, the vehicle is able to get out of the atmosphere, but without enough fuel to complete the mission.

- Figures 10 (HEO high altitude) and 14 (HEO low altitude) show that the aeroassisted maneuver is much more convenient if the HEO altitude is higher. Lowering the initial altitude reduces the convenience but considerably extends the range of feasibility. The reduced convenience at low altitudes is due to the fact that the Hohmann transfer requires a lower fuel quantity with the lowered the initial altitude for the same final LEO. In contrast, for the corresponding aeroassisted maneuver, even if the TPS is less consistent, the propulsive contribution remains significant.

- Figures 10-13, considered in pairs (the first two and the second two), provide a performance comparison between PICA and AVCOAT, respectively, for high initial orbits in the case of low and high aerodynamic efficiency. The comparison is in favor of the use of PICA, albeit less marked than for GEO orbits. For high orbits for a vehicle with high ε, the performance of PICA remains considerable; however, the performance of AVCOAT, although lower than PICA, is notable. In this context, it seems appropriate to comment on NASA's recent evaluation concerning the choice of material for ablative TPS of the Orion spacecraft. The final decision was between PICA and AVCOAT, and NASA ultimately chose the latter. PICA has been used to date only on a re-entry capsule of the Stardust probe, while AVCOAT has a long history of reliability (Apollo missions). Though the results of the present work provide evidence of the tangible prevalence of the PICA performance, the reliability of AVCOAT most likely outweighs these performance considerations.

- The increase in aerodynamic efficiency has less impact in the case of lower orbits (Figures 14 and 15).

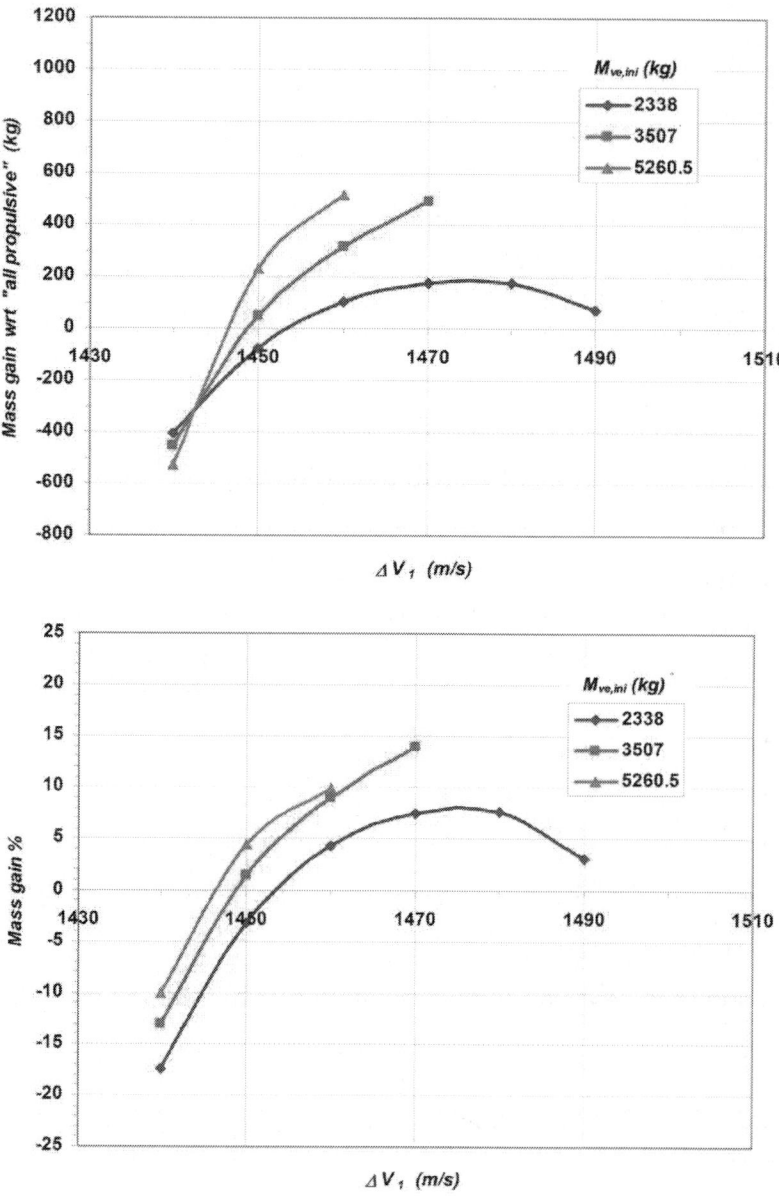

Figure 10: HEO-LEO scenario 7: HEO altitude influence; reference case for HEOH.

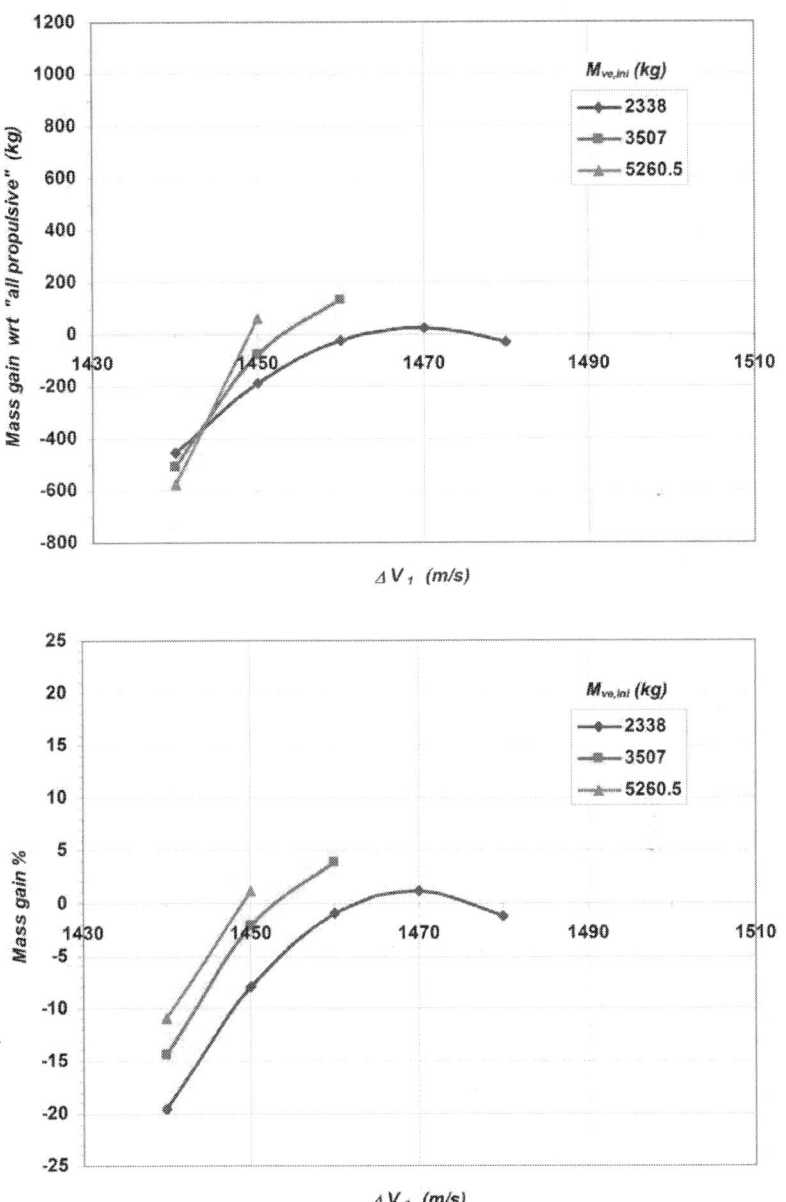

Figure 11: HEO-LEO scenario 8: material (AVCOAT) influence w.r.t. HEOH.

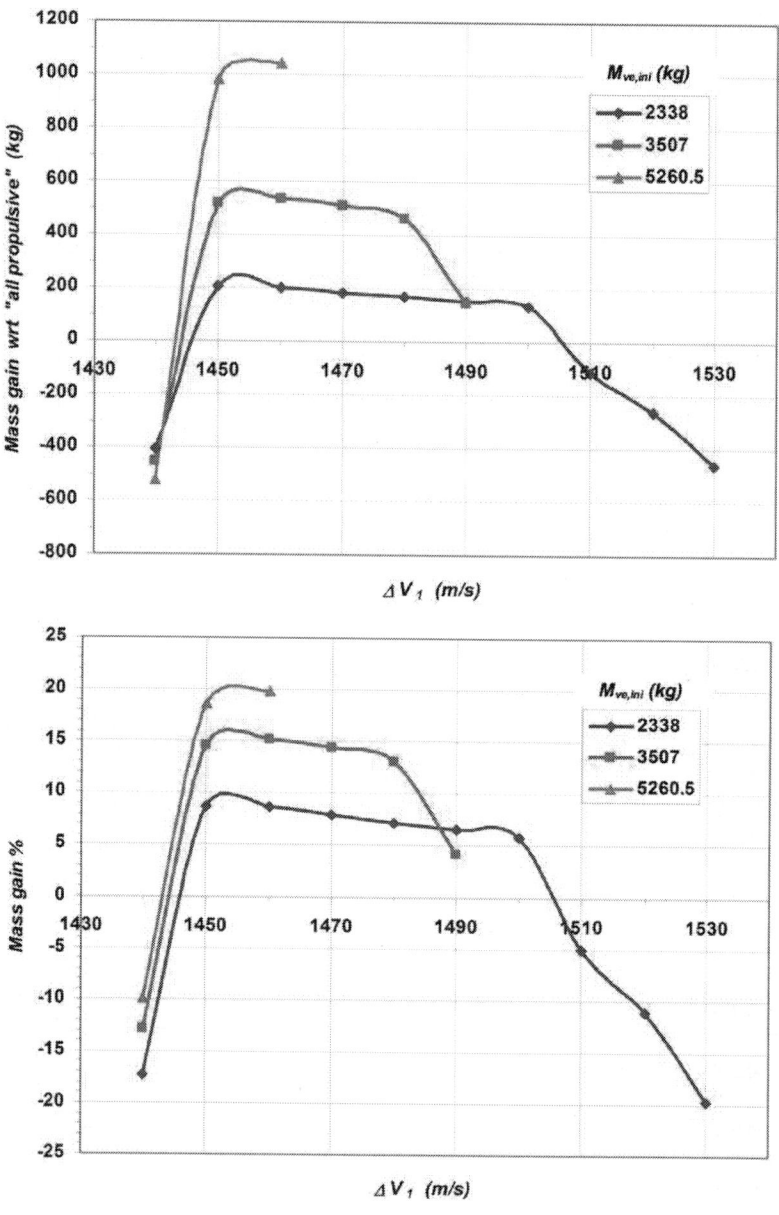

Figure 12: HEO-LEO scenario 9: e influence w.r.t. HEOH.

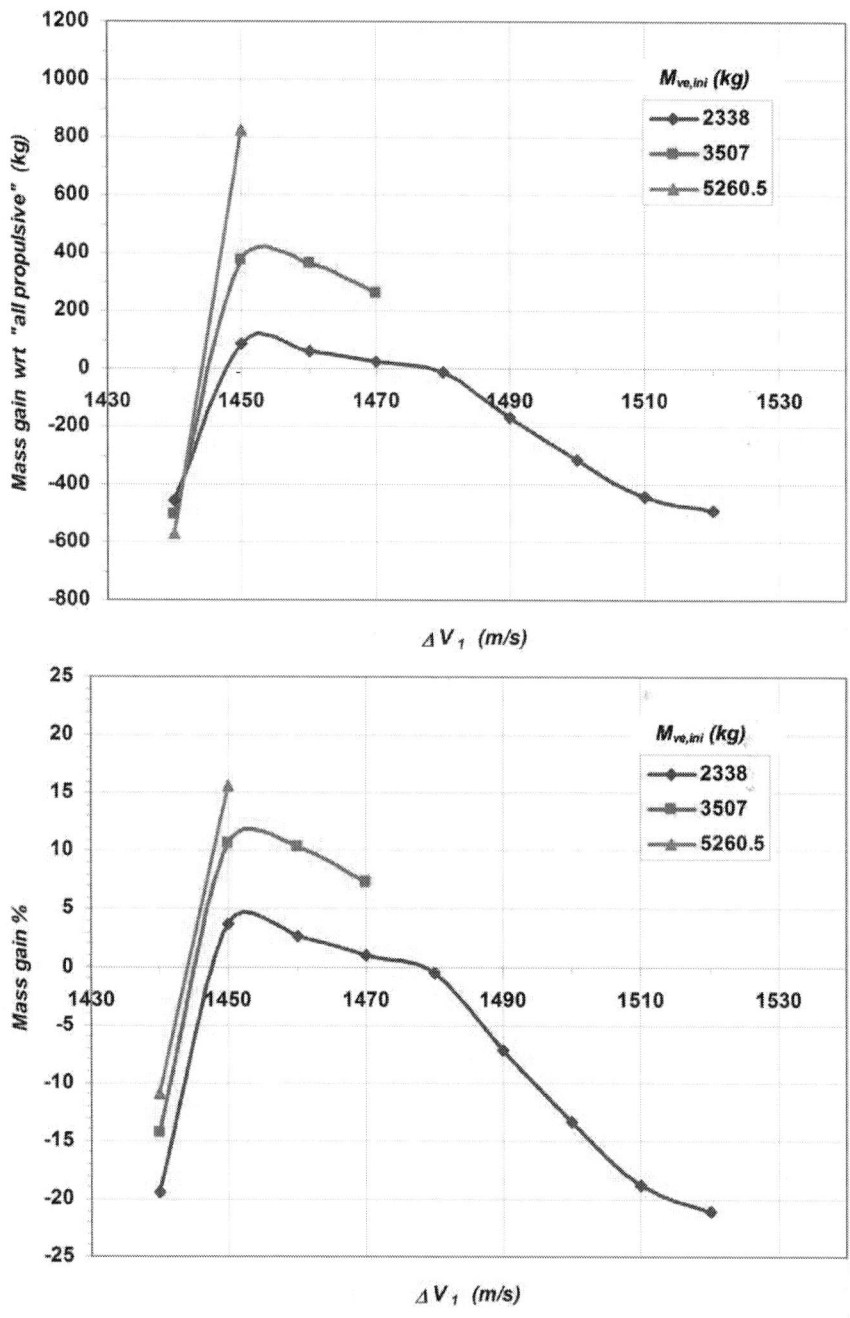

Figure 13: HEO-LEO scenario 10: Material (AVCOAT) + e influence w.r.t. HEOH.

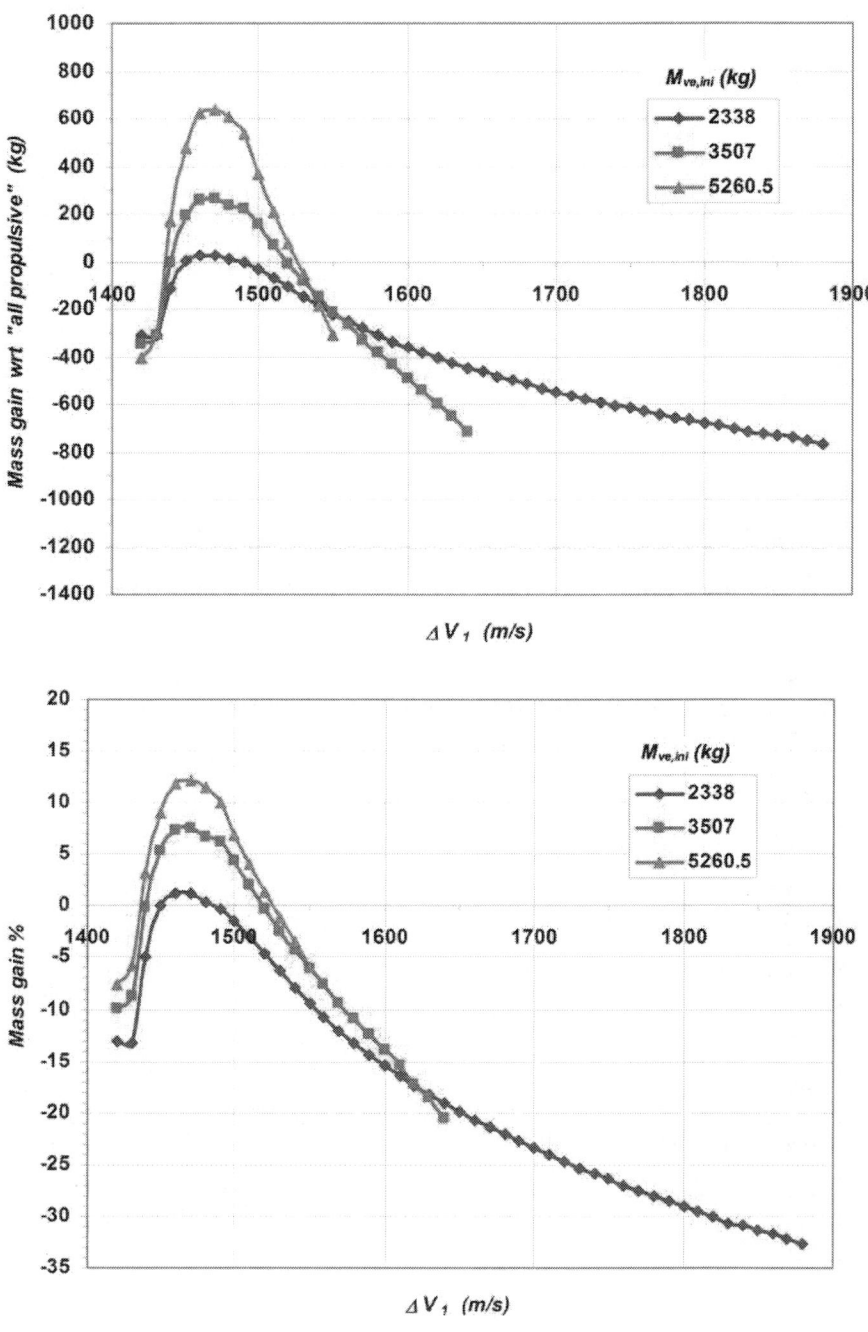

Figure 14: HEO-LEO scenario 11: HEO altitude influence; reference case for HEOL.

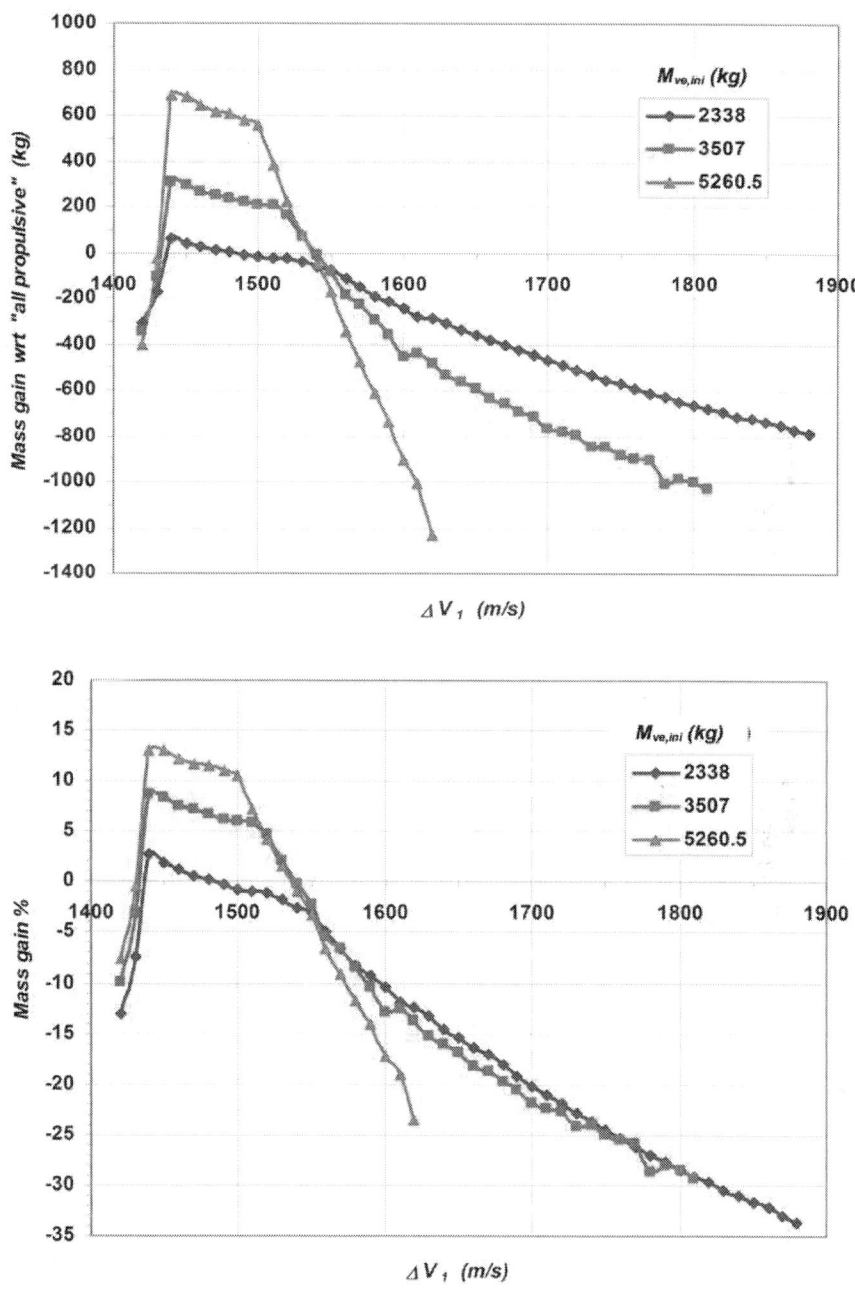

Figure 15: HEO-LEO scenario 12: e influence w.r.t. HEOL.

SUMMARY AND CONCLUSIONS

Some aeroassisted transfers between coplanar orbits were analyzed to examine the influence of the aerodynamic characteristics of a spacecraft and the characteristics of its heat shield on the maneuver's final performance.

The study was conducted with an original tool developed by the authors comprising highly representative models for thermal analysis, atmospheric flight dynamics, and an optimizer based on a genetic algorithm. The convenience of the maneuver was evaluated by comparison with the corresponding extra-atmospheric Hohmann transfer with two impulses.

The results of the case study indicated that there is a range of feasibility for the maneuver, depending on the first propulsive impulse of deorbit, within which there is, in turn, an interval in which the maneuver is more convenient than the homologous Hohmann transfer in terms of the total mass savings achieved (TPS and propellant). In particular, the results show that in terms of the convenience of the aeroassisted maneuver, the influence of the ablative material type used (low-density advantage) and of the aerodynamic efficiency is the most relevant, whereas other characteristics play a minor role. Significant benefits can be obtained from a careful thermal design, with a suitable thermal safety factor and high-performance adhesives for the bond-line.

For all cases, the highest convenience, and consequently the corresponding optimal trajectories, is obtained when it is not necessary to apply the second propulsion impulse. The most important future development for this type of analysis is increasing the number of cases analyzed in terms of materials, types of maneuvers and spacecraft characteristics.

REFERENCES

1. A. Mazzaracchio and M. Marchetti, "A Probabilistic Sizing Tool and Monte Carlo Analysis for Entry Vehicle Ablative Thermal Protection Systems," Acta Astronautica, Vol. 66, No. 5-6, 2010, pp. 821-835. doi:10.1016/j.actaastro.2009.08.033

2. A. Mazzaracchio and M. Marchetti, "Coupled Aeroassisted Orbital Plane Change Manoeuvre and Thermal Protection System Optimisation," 61st International Astronautical Congress, Prague, September 27-October 1, 2010.

3. P. Charbonneau and B. Knap, "A User'S Guide to Pikaia 1.0," Boulder, Colorado, 1995.

4. P. Charbonneau, "An Introduction to Genetic Algorithms for Numerical Optimization," Boulder, Colorado, 2002.

5. P. Charbonneau, "Release Notes for Pikaia 1.2," Boulder, Colorado, 2002.

6. C. Gogu, T. Matsumura, R. T. Haftka and A. V. Rao, "Aeroassisted Orbital Transfer Trajectory Optimization Considering Thermal Protection System Mass," Journal of Guidance, Control and Dynamics, Vol. 32, No. 3, 2009, pp. 927-938.doi:10.2514/1.37684

7. NASA Marshall Space Flight Center, "X-37 Demonstrator to Test Future Launch Technologies in Orbit and Reentry Environments," NASA Facts, May 2003, FS-2003-05- 65-MSFC.

8. Y. Y. Shi and D. H. Young, "Minimum Fuel Coplanar Aeroassisted Orbital Transfer Using Collocation and Nonlinear Programming," Flight Mechanics/Estimation Theory Symposium, NASA Goddard Space Flight Center, 1991, pp. 461-480 (SEEN92-1407005-13).

9. US Standard Atmosphere, US Government Printing Office, Washington DC, 1976.

10. T. J. Collins, W. M. Congdon, S. S. Smeltzer and K. S. Whitley, "High-Temperature Structures, Adhesives, and Advanced Thermal Protection Materials for Next-Generation Aeroshell Design," NASA Langley Research Center, 2006, Paper 2M-02-2005.

NOMENCLATURE

C_{D0} zero-lift drag coefficient

$C_{L,\alpha}$ derivative of C_L w.r.t. the angle of attack

$C_{L,\max}$ maximum lift coefficient

ff fitness function

g_0 gravitational acceleration at sea level (m/s²)

H_A altitude of the initial LEO (m)

H_{atm} altitude of the sensible atmosphere (m)

H_B altitude of the final LEO (m)

I_{sp} propellant specific impulse (s)

K_D induced drag factor

l_{ve} vehicle length (m)

$m_{TPS,los}$ TPS mass lost during atmospheric pass (kg)

$m_{ve,fin}$ final vehicle mass (kg)

$m_{ve,fin,ap}$ "all propulsive" case final vehicle mass (kg)

$m_{ve,i}$ vehicle mass at atmospheric entry (kg)

$m_{ve,ini}$ initial vehicle mass (kg)

$m_{ve,u}$ vehicle mass at atmospheric exit (kg)

R_A initial HEO radius (m)

R_{atm} radius of the sensible atmosphere (m)

R_B final LEO radius (m)

$R_{f,'j'}$ reward factor for component "j"

$R_{\oplus}$ Earth's radius (m)

r_b vehicle body radius (m)

S vehicle reference surface (m²)

$S_{TPS,ve}$ vehicle TPS total surface (m²)

T_{BL} bond-line temperature (K)

$T_{BL,lim}$ bond-line limit temperature (K)

TSF thermal safety factor

V velocity modulus (m/s)

V_A circular orbit speed in initial HEO (m/s)

V_B circular orbit speed in final LEO (m/s)

V_i speed at atmospheric entry (m/s)

V_u speed at atmospheric exit (m/s)

$w_{,j}$ multiplicative weight of component "j"

wc_{ve} vehicle wing cord (m)

ws_{ve} vehicle wing span (m)

ΔV speed variation, impulse (m/s)

ΔV_A first impulse for Hohmann transfer (m/s)

ΔV_B second impulse for Hohmann transfer (m/s)

ΔV_{AP} "all propulsive" (impulse) speed variation (m/s)

ΔV_1 deorbit (impulse) speed variation (m/s)

$\Delta V_{L,min}$ minimum deorbit impulse (m/s)

ΔV_2 boost (impulse) speed variation (m/s)

ΔV_3 circularizing (impulse) speed variation (m/s)

α angle of attack (rad)

ε aerodynamic efficiency

γ_i flight path angle at atmospheric entry (rad)

γ_u flight path angle at atmospheric exit (rad)

μ gravitational parameter (m³/s²)

σ bank angle (rad)

Subscripts

A value at initial HEO

ap "all propulsive"

B value at final LEO

BL bond-line

HF heat flux

i value at atmospheric entry

los lost

m mass

TPS Thermal Protection System

u value at atmospheric exit

ve vehicle

CITATION

A. Mazzaracchio and M. Marchetti, "Effect of Spacecraft Aerodynamics and Heat Shield Characteristics on Optimal Aeroassisted Transfer," Engineering, Vol. 4 No. 6, 2012, pp. 307-320. doi: 10.4236/eng.2012.46040.

CHAPTER 2

Measurements of Local Heat Flux and Water-Side Heat Transfer Coefficient In Water Wall Tubes

Jan Taler[1] and Dawid Taler[2]

[1] Department of Thermal Power Engineering, Cracow University of Technology, Cracow, Poland

[2] Institute of Heat Transfer Engineering and Air Protection, Cracow University of Technology, Cracow, Poland

INTRODUCTION

Measurements of heat flux and heat transfer coefficient are subject of many current studies. A proper understanding of combustion and heat transfer in furnaces and heat exchange on the water-steam side in water walls requires accurate measurement of heat flux which is absorbed by membrane furnace walls. There are three broad categories of heat flux measurements of the boiler water-walls: (1) portable heat flux meters inserted in inspection ports [1], (2) Gardon type heat flux meters welded to the sections of the boiler tubes [1-4],(3) tubular type instruments placed between two adjacent boiler tubes [5-14]. Tubular type and Gardon meters strategically placed on the furnace tube wall can be a valuable boiler diagnostic device for monitoring of slag deposition. If a heat flux instrument is to measure the absorbed heat flux correctly, it must resemble the boiler tube as closely as possible so far as radiant heat exchange with the flame and surrounding surfaces is concerned. Two main factors in this respect are the emissivity and the temperature of the absorbing surface, but since the instrument will almost always be coated with ash, it is generally the properties of the ash and not the instrument that dominate the situation. Unfortunately, the thermal conductivity can vary widely. Therefore, accurate measurements will only be performed if the deposit on the meter is representative of that on the surrounding tubes. The tubular type instruments known also as flux-tubes meet this requirement. In these devices the measured boiler tube wall temperatures are used for the evaluation of the heat fluxq_m. The measuring tube is fitted with two thermocouples in holes of known radial spacing r_1 and r_2. The thermocouples are led away to the junction box where they are connected differentially to give a flux related electromotive force.

The use of the one dimensional heat conduction equation for determining temperature distribution in the tube wall leads to the simple formula

$$q_m = \frac{k\left(f_1 - f_2\right)}{r_0 \ln\left(r_1/r_2\right)}. \tag{1}$$

The accuracy of this equation is very low because of the circumferential heat conduction in the tube wall.

However, the measurement of the heat flux absorbed by water-walls with satisfactory accuracy is a challenging task. Considerable work has been done in recent years in this field.Previous attempts to accurately measure the local heat flux to membrane water walls in steam boilers failed due to calculation of inside heat transfer coefficients. The heat flux can only be determined accurately if the inside heat transfer coefficient is measured experimentally.

New numerical methods for determining the heat flux in boiler furnaces, based on experimentally acquired interior flux-tube temperatures, will be presented. The tubular type instruments have been designed to provide a very accurate measurement of absorbed heat flux q_m, inside heat transfer coefficient h_{in}, and water steam temperature T_f.

Two different tubular type instruments (flux tubes) were developed to identify boundary conditions in water wall tubes of steam boilers.

The first meter is constructed from a short length of eccentric bare tube containing four thermocouples on the fire side below the inner and outer surfaces of the tube. The fifth thermocouple is located at the rear of the tube on the casing side of the water wall tube. First, formulas for the view factor defining the heat flux distribution at the outer surface of the flux tube were derived. The exact analytical expressions for the view factor compare very well with approximate methods for determining view factor which are used by the ANSYS software. The meter is constructed from a short length of eccentric tube containing four thermocouples on the fireside below the inner and outer surfaces of the tube. The fifth thermocouple is located at the rear of the tube (on the casing side of the water-wall tube). The boundary conditions on the outer and inner surfaces of the water flux-tube must then be determined from temperature measurements at the interior locations. Four K-type sheathed thermocouples, 1 mm in diameter, are inserted into holes, which are parallel to the tube axis. The thermal conduction effect at the hot junction is minimized because the thermocouples pass through isothermal holes. The thermocouples are brought to the rear of the tube in the slot machined in the tube wall. An austenitic cover plate with the thickness of 3 mm – welded to the tube – is used to protect the thermocouples from the incident flame radiation. A K-type sheathed thermocouple with a pad is used to measure the temperature at the rear of the flux-tube. This temperature is almost the same as the water-steam temperature.

The non-linear least squares problem was solved numerically using the Levenberg–Marquardt method. The temperature distribution at the cross section of the flux tube was determined at every iteration step using the method of separation of variables. The heat transfer conditions in adjacent boiler tubes have no impact on the temperature distribution in the flux tubes.

The second flux tube has two longitudinal fins which are welded to the eccentric bare tube. In contrast to existing devices, in the developed flux-tube fins are not welded to adjacent water-wall tubes. Temperature distribution in the flux-tube is symmetric and not disturbed by different temperature fields in neighboring tubes. The temperature dependent thermal conductivity of the flux-tube material was assumed. An inverse problem of heat conduction was solved using the least squares method. Three unknown parameters were estimated using the Levenberg-Marquardt method. At every iteration step, the temperature distribution over the cross-section of the heat flux meter was computed using the ANSYS CFX software. Test calculations were carried out to assess accuracy of the presented method. The uncertainty in determined parameters was calculated using the variance propagation rule by Gauss. The presented method is appropriate for membrane water-walls.

The developed meters have one particular advantage over the existing flux tubes to date. The temperature distribution in the flux tube is not affected by the water wall tubes, since the flux tube is not connected to adjacent water wall tubes with metal bars, referred to as membrane or webs. To determine the unknown parameters only the temperature distribution at the cross section of the flux tube must be analysed.

TUBULAR TYPE HEAT FLUX METER MADE OF A BARE TUBE

Heat flux meters are used for monitoring local waterwall slagging in coal and biomass fired steam boilers [5-19].

The tubular type instruments (flux tubes) [10-14,19] and other measuring devices [15-18] were developed to identify boundary conditions in water wall tubes of steam boilers. The meter is constructed from a short length of eccentric tube containing four thermocouples on the fire side below the inner and outer surfaces of the tube. The fifth thermocouple is located at the rear of the tube on the casing side of the water wall tube.

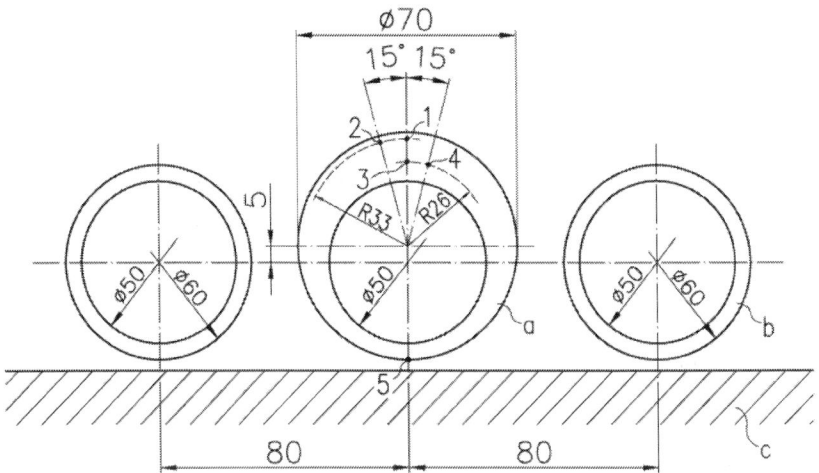

Figure 1: The heat flux tube placed between two water wall tubes, a – flux tube, b – water wall tube, c – thermal insulation.

The boundary conditions at the outer and inner surfaces of the water flux-tube must then be determined from temperature measurements at the interior locations. Four K-type sheathed thermocouples, 1 mm in diameter, are inserted into holes, which are parallel to the tube axis. The thermal conduction effect at the hot junction is minimized because the thermocouples pass through isothermal holes. The thermocouples are brought to the rear of the tube in the slot machined in the protecting pad. An austenitic cover plate with the thickness of 3 mm welded to the tube is used to protect the thermocouples from the incident flame radiation. A K-type sheathed thermocouple with a pad is used to measure the temperature at the rear of the flux-tube. This temperature is almost the same as the water-steam temperature. A method for determining fireside heat flux, heat transfer coefficient on the inner surface and temperature of water-steam mixture in water-wall tubes is developed. The unknown parameters are estimated based on the temperature measurements at a few internal locations from the solution of the inverse heat conduction problem. The non-linear least squares problem is solved numerically using the Levenberg–Marquardt method. The diameter of the measuring tube can be larger than the water-wall tube diameter. The view factor defining the distribution of the heat flux on the measuring tube circumference is determined using exact analytical formulas and compared with the results obtained numerically using ANSYS software. The method developed can also be used for an assessment of scale deposition on the inner surfaces of the water wall tubes or slagging on the fire side. The presented method is suitable for water walls made of bare tubes as well as for membrane water walls. The heat transfer conditions in adjacent boiler tubes have no impact on the temperature distribution in the flux tubes.

View Factor for Radiation Heat Transfer between Heat Flux Tube and Flame

The heat flux distribution in the flux tube depends heavily on the heat flux distribution on its outer surface. To determine the heat flux distribution q as a function of angular coordinate φ, the analytical formulas for the view factor ψ, defining radiation interchange between an infinitesimal surface on the outer flux tube circumference and the infinite flame or boiler surface, will be derived. The heat flux absorbed by the outer surface of the heat flux tube $q(\varphi)$ is given by

$$q(\varphi) = q_m \psi(\varphi). \tag{2}$$

The specific thermal load of the water wall q_m is defined as the ratio of the heat transfer rate absorbed by the waterwall to the projected surface area of the water wall. The view factor is the fraction of the radiation leaving the surface element located on the flux tube surface that arrives at the flame surface. The view factor can be computed from

$$\psi = \frac{1}{2}\left(\sin\delta_1 + \sin\delta_2\right). \tag{3}$$

The angles δ_1 and δ_2 are formed by the normal to the flux tube at φ and the tangents to the flux tube and adjacent water-wall tube (Figures 2,4,6). Positive values of δ_1 are measured clockwise with respect to the normal while positive values of δ_2 are measured counterclockwise with respect to the normal. The radial coordinate ro of the flux tube outer surface measured from the center 0 (Figure 2) is

$$r_o = e\cos\varphi + \sqrt{b^2 - e^2(\sin\varphi)^2}. \tag{4}$$

where: e– eccentric (Figure 2), b– outer radius of flux-tube.

The angleφ1can be expressed in terms of the angle φ, flux tube outer radius b, and eccentric e (Figure 2)

$$\varphi_1 = \arcsin\left[\frac{\left(e\cos\varphi + \sqrt{b^2 - (e\sin\varphi)^2}\right)\sin\varphi}{b}\right], \quad \varphi_1 \le \frac{\pi}{2}, \tag{5}$$

$$\varphi_1 = \pi - \arcsin\left[\frac{\left(e\cos\varphi + \sqrt{b^2 - (e\sin\varphi)^2}\right)\sin\varphi}{b}\right], \quad \frac{\pi}{2} \le \varphi_1 \le \pi. \tag{6}$$

First, the view factor for the angle interval $0 \le \varphi_1 \le \varphi_{1,11}$ was determined

$$\psi = \frac{1 + \cos\varphi_1}{2}, \quad 0 \le \varphi_1 \le \varphi_{1,11}. \tag{7}$$

The limit angle $\varphi_{1,11}$(Figures 2 and 3) is given by

$$\varphi_{1,11} = \arccos\frac{c-e}{b}, \tag{8}$$

where c is the outer radius of the boiler tube.

Next the view factor in the angle interval $\varphi_{1,11} \le \varphi_1 \le \varphi_{1,12}$ will be determined. The limit angle $\varphi_{1,12}$ is: $\varphi_{1,12} = \varphi_1(\varphi = \pi/2) = (\pi/2) + \arcsin(e/b)$ (Figure3). The view factor ψ is computed from Eq. (2), taking into account that (Figure 4)

$$\delta_1 = \frac{\pi}{2}, \ \delta_2 = \frac{\pi}{2} - (\varepsilon + \varphi_1), \ \varepsilon = \beta + \gamma - \frac{\pi}{2}, \ x_i = b\sin\varphi_1, \ x_i = b\cos\varphi_1,$$

$$\beta = \arcsin\frac{c}{\sqrt{(t-x_i)^2 + (y_i + e)^2}}, \ \gamma = \arcsin\frac{t-x_i}{\sqrt{(t-x_i)^2 + (y_i + e)^2}}, \ \varphi_{11} \le \varphi_1 \le \varphi_{12}, \tag{9}$$

where t is the pitch of the water wall tubes.

Next the view factor ψ (φ) is determined in the angle interval $\varphi_{1,12} \le \varphi_1 \le \varphi_{1,13}$ (Figures 3 and5).

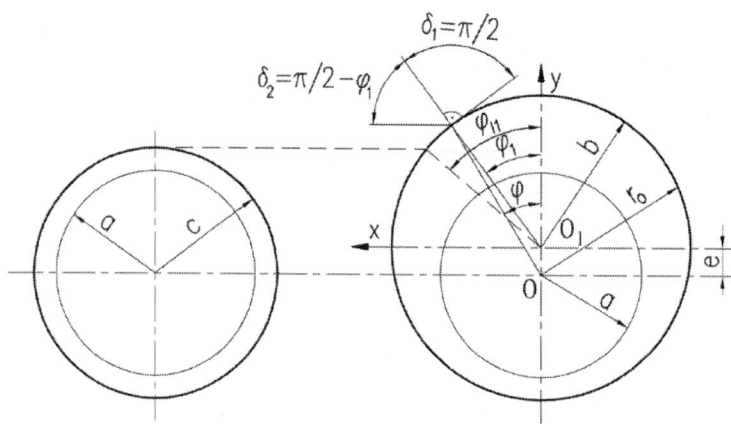

Figure 2: Determination of view factor in the angle interval $0 \le \varphi_1 \le \varphi_{1,11}$

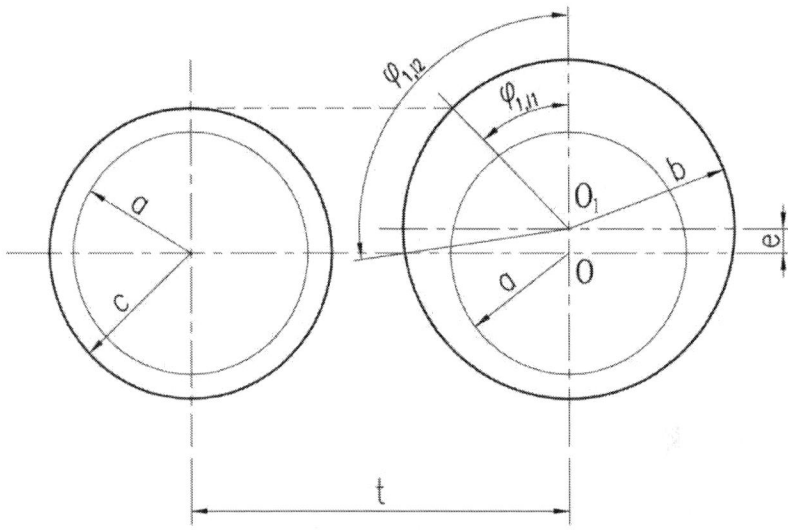

Figure 3: Limit angles $\varphi_{1,l1}$ and $\varphi_{1,l2}$

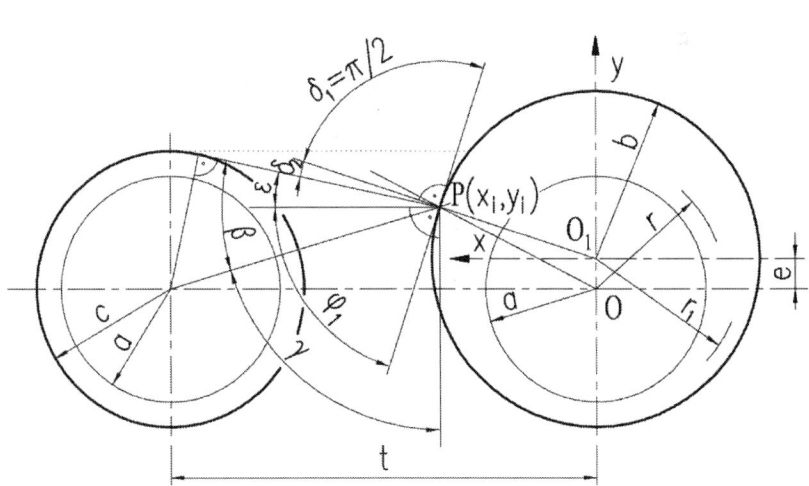

Figure 4: Determination of view factor in the angle interval $\varphi_{1,l1} \leq \varphi_1 \leq \varphi_{1,l2}$

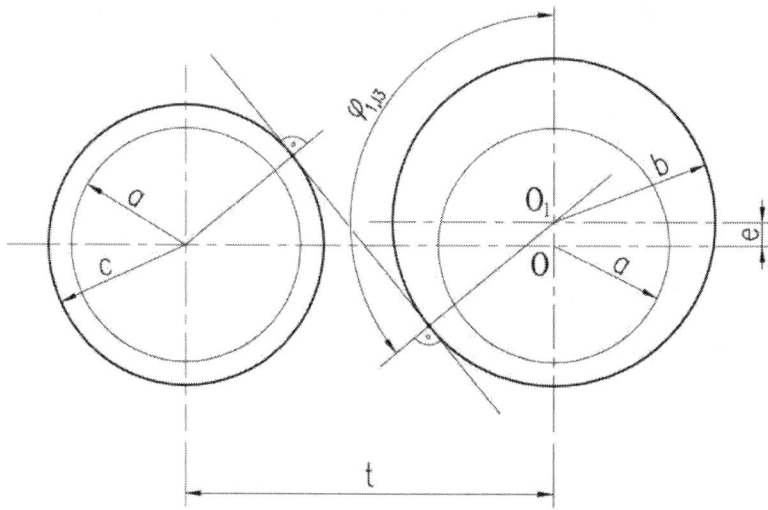

Figure 5: Limit angle $\varphi_{1,13}$

The limit angle $\varphi_{1,13}$ (Figure 5) can be expressed as

$$\varphi_{1,13} = \frac{\pi}{2} + \omega + \kappa,$$ (10)

where the angles κ i ω are given by

$$\kappa = \arctan \frac{b-c}{t},$$ (11)

$$\omega = \arccos \frac{b+c}{\sqrt{t^2 + e^2}}.$$ (12)

The view factor ψ in the interval $\varphi_{1,12} \leq \varphi_1 \leq \varphi_{1,13}$ is calculated from the following expression (Figure 6)

$$\psi = \frac{1}{2}\left(\sin\delta_2 - \sin\delta_1\right), \varphi_{1,12} \leq \varphi_1 \leq \varphi_{1,13}, \tag{13}$$

Where

$$\delta_1 = \frac{\pi}{2}, \tag{14}$$

$$\delta_2 = \varepsilon + \varphi_1 - \frac{\pi}{2}, \tag{15}$$

$$\varepsilon = \beta + \gamma - \frac{\pi}{2} \tag{16}$$

$$\beta = \arcsin\frac{c}{\sqrt{\left(t - x_i\right)^2 + \left(y_i + e\right)^2}}, \tag{17}$$

$$\gamma = \pi - \arcsin\frac{t - x_i}{\sqrt{\left(t - x_i\right)^2 + \left(y_i + e\right)^2}}, \tag{18}$$

$$x_i = b\sin\varphi_1, \tag{19}$$

$$y_i = b\cos\varphi_1. \tag{20}$$

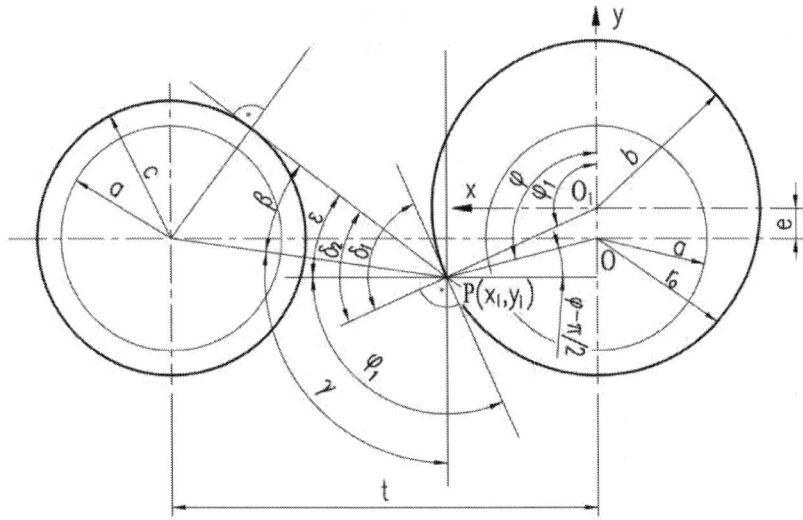

Figure 6: Determination of view factor in the angle interval $\varphi_{1,12} \leq \varphi_1 \leq \varphi_{1,13}$

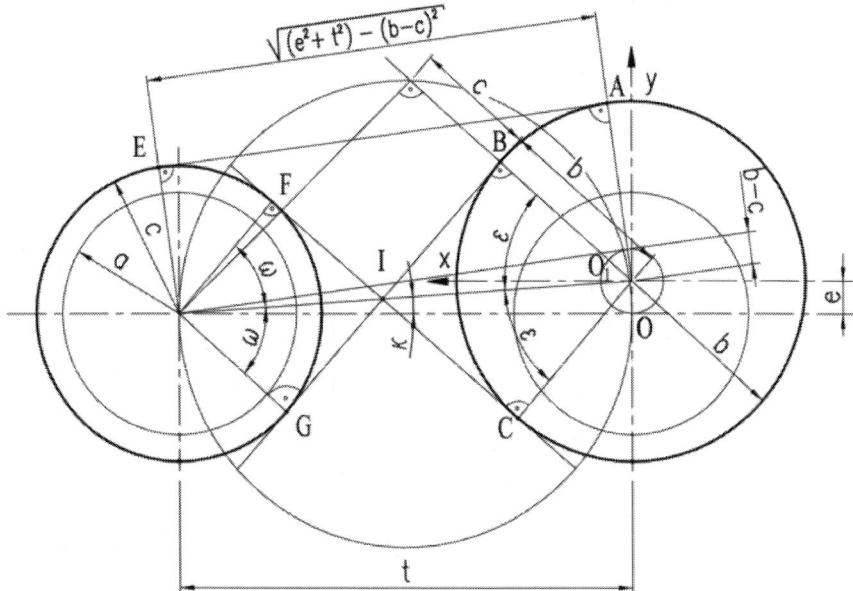

Figure 7: Determination of mean view factor ψ_{bs} for boiler setting over tube pitch t using the crossed string method

Radiation leaving the flame reaches also the boiler setting. The view factor for the radiation heat exchange between boiler setting and rear side of the measuring tube can be calculated in similar way as for the forward part. The mean heat flux q_{bs} resulting from the radiation heat transfer between the flame and the boiler setting can be determined using the crossed-string method [20-21].

The mean value of the view factor ψbs over the pitch length t is calculated from (Figure7)

$$\psi_{bs} = \frac{1}{2t}\left[\left(FC + BG\right) - \left(FG + BC\right)\right]$$

(21)

After substituting the lengths of straight FC and BG and circular segments FG and BC into Eq. (21), the mean value of the view factor over the boiler setting can be expressed as:

$$\psi_{bs} = \frac{b+c}{t}\left(\tan\omega - \omega\right).$$

(22)

The mean heat flux over the setting surface is

$$q_{bs} = q_m \psi_{bs}.$$

(23)

The angle ω is determined from

$$\tan\omega = \frac{\sqrt{e^2 + t^2 - \left(b + c\right)^2}}{b + c},$$

(24)

If the diameters of the heat flux and waterwall tubes are equal, then Eq. (24) simplifies to

$$\tan \omega = \sqrt{\left(\frac{t}{2c}\right)^2 - 1}. \tag{25}$$

The view factor for the radiation heat exchange between boiler setting and rear side of the measuring tube can be calculated in similar way as for the forward part. The view factor in the angle interval $\varphi_{1,14} \leq \varphi_1 \leq \varphi_{1,15}$ (Figure 8), accounting for the setting radiation, is given by

$$\psi = \psi_{bs} \cdot \frac{1}{2}\left(\sin \delta_2 - \sin \delta_1\right), \ \varphi_{1,14} \leq \varphi_1 \leq \varphi_{1,15} \tag{26}$$

where the limit angle φ_1,

$$\psi_{1,14} = \frac{\pi}{2} - \omega + \kappa. \tag{27}$$

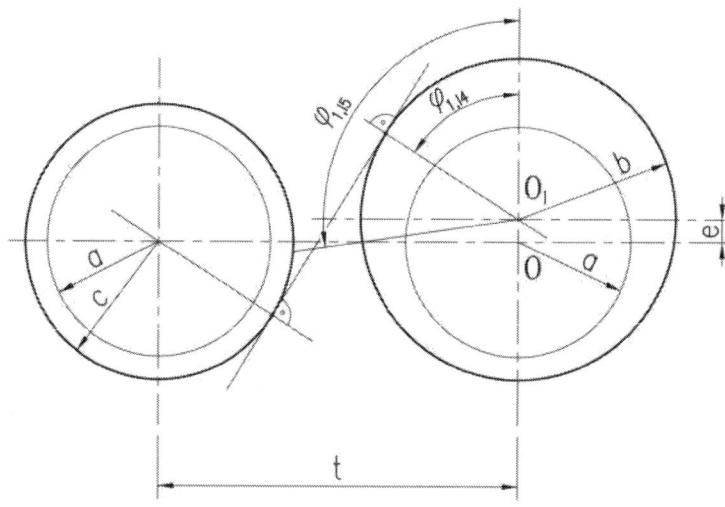

Figure 8: Limit angles $\varphi_{1,14}$ and $\varphi_{1,15} = \varphi_{1,12} = \left(\pi / 2\right) + \arcsin\left(e / b\right)$

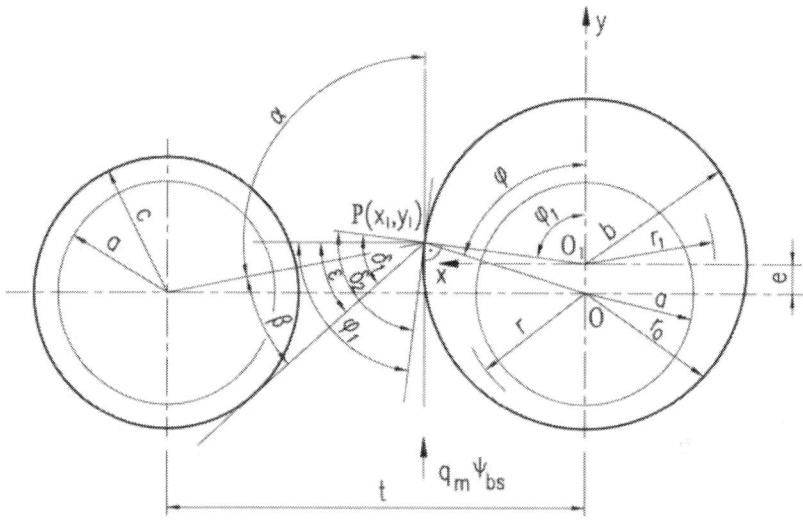

Figure 9: Determination of view factor in the angle interval $\varphi_{1,14} \leq \varphi_1 \leq \varphi_{1,15}$

The angles δ_1 and δ_2 are (Figure 9)

$$\delta_1 = \frac{\pi}{2} + \varepsilon - \varphi_1,\qquad\qquad(28)$$

$$\delta_2 = \frac{\pi}{2},\qquad\qquad(29)$$

Where

$$\varepsilon = \beta + \gamma - \frac{\pi}{2}, \tag{30}$$

$$\beta = \arcsin \frac{c}{\sqrt{(t - x_i)^2 + (y_i + e)^2}}, \tag{31}$$

$$\gamma = \pi - \arcsin \frac{t - x_i}{\sqrt{(t - x_i)^2 + (y_i + e)^2}}, \tag{32}$$

$$x_i = b \sin \varphi_1, \tag{33}$$

$$y_i = b \cos \varphi_1. \tag{34}$$

The view factor ψ in the interval $\varphi_{1, l5} \le \varphi \le \pi$, where $\varphi_{1, l5} = \varphi_{1, l2}$, is given by

$$\psi = \psi_{bs} \cdot \frac{1}{2} \left(\sin \delta_1 + \sin \delta_2 \right), \varphi_{1, l5} \le \varphi \le \pi, \tag{35}$$

Where

$$\delta_1 = \varphi_1 - \varepsilon - \frac{\pi}{2}, \tag{36}$$

$$\delta_2 = \frac{\pi}{2}, \tag{37}$$

$$\varepsilon = \frac{\pi}{2} - \left(\gamma - \beta \right), \tag{38}$$

$$\beta = \arcsin \frac{c}{\sqrt{(t - x_i)^2 + (y_i + e)^2}}, \tag{39}$$

$$\gamma = \pi - \arcsin \frac{t - x_i}{\sqrt{(t - x_i)^2 + (y_i + e)^2}}, \tag{40}$$

$$x_i = b\cos\left(\varphi_1 - \frac{\pi}{2}\right), \tag{41}$$

$$y_i = -b\sin\left(\varphi_1 - \frac{\pi}{2}\right). \tag{42}$$

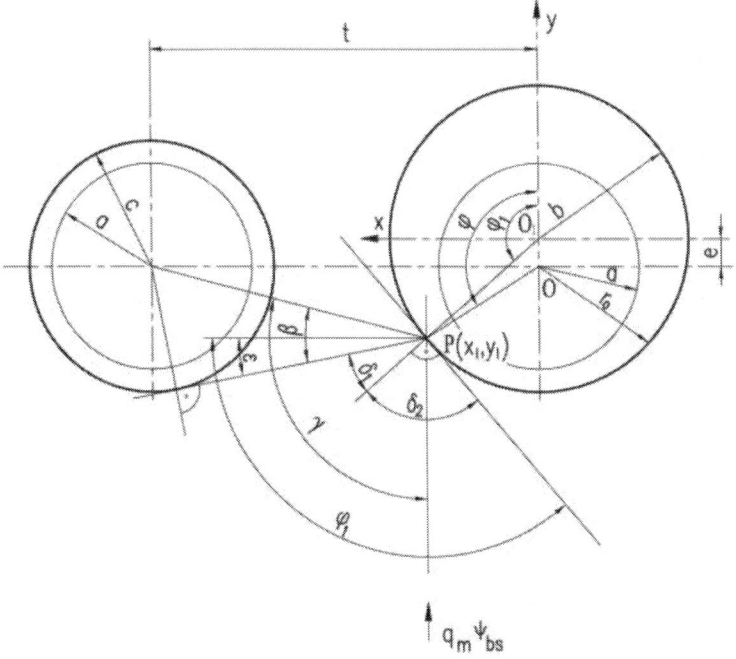

Figure 10: Determination of view factor in the angle interval $\varphi_{1,15} \le \varphi_1 \le \pi$

The total view factor accounts for the radiation heat exchange between the heat flux tube and flame and between the heat flux tube and the boiler setting.

Theory of the Inverse Problem

At first, the temperature distribution at the cross section of the measuring tube will be determined, i.e. the direct problem will be solved. Linear direct heat conduction problem can be solved using an analytical method. The temperature distribution will also be calculated numerically using the finite element method (FEM). In order to show accuracy of a numerical approach, the results obtained from numerical and analytical methods will be compared. The following assumptions have been made:

- thermal conductivity of the flux tube material is constant,
- heat transfer coefficient at the inner surface of the measuring tube does not vary on the tube circumference,
- rear side of the water wall, including the measuring tube, is thermally insulated,
- diameter of the eccentric flux tube is larger than the diameter of the water wall tubes,
- the outside surface of the measuring flux tube is irradiated by the flame, so the heat absorption on the tube fire side is non-uniform.

The cylindrical coordinate system is shown in Figure11.

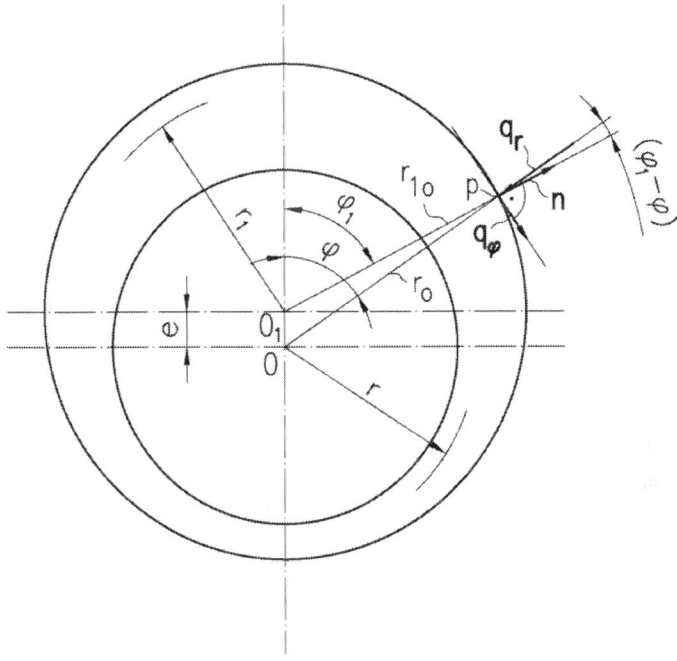

Figure 11: Approximation of the boundary condition on the outer tube surface.

The temperature distribution in the eccentric heat flux tube is governed by heat conduction

$$\frac{1}{r}\frac{\partial}{\partial r}\left(kr\frac{\partial\theta}{\partial r}\right)+\frac{1}{r}\frac{\partial}{\partial\varphi}\left(\frac{k}{r}\frac{\partial\theta}{\partial\varphi}\right)=0 \tag{43}$$

subject to the following boundary conditions

$$k\nabla\theta\cdot\mathbf{n}\Big|_{r=r_o}=q_m\psi\left(\varphi\right) \tag{44}$$

$$k\frac{\partial\theta}{\partial r}\bigg|_{r=a}=h\theta\big|_{r=a} \tag{45}$$

The left side of Eq. (44) can be transformed as follows (Figure11)

$$k \nabla \theta \cdot \mathbf{n}\big|_{r=r_o} = \left(\mathbf{q}_r + \mathbf{q}_\varphi\right) \cdot \mathbf{n}\big|_{r=r_o} =$$

$$= \left[k \frac{\partial T}{\partial r} \cos\left(\varphi_1 - \varphi\right) + \frac{k}{r} \frac{\partial T}{\partial \varphi} \sin\left(\varphi_1 - \varphi\right) \right]_{r=r_o} \tag{46}$$

The second term in Eq. (46) can be neglected since it is very small and the boundary condition (44) simplifies to

$$k \frac{\partial \theta}{\partial r}\bigg|_{r=r_o} = \frac{q_m \psi\left(\varphi\right)}{\cos\left(\varphi_1 - \varphi\right)} \tag{47}$$

The heat flux over the tube circumference can be approximated by the Fourier polynomial

$$\frac{q_m \psi\left(\varphi\right)}{\cos\left(\varphi_1 - \varphi\right)} = q_0 + \sum_{n=1}^{\infty} q_n \cos\left(n\varphi\right) \tag{48}$$

Where

$$q_0 = \frac{1}{\pi} \int_0^\pi \frac{q_m \psi\left(\varphi\right)}{\cos\left(\varphi_1 - \varphi\right)} d\varphi,$$

$$q_n = \frac{2}{\pi} \int_0^\pi \frac{q_m \psi\left(\varphi\right)}{\cos\left(\varphi_1 - \varphi\right)} \cos\left(n\varphi\right) d\varphi, \ n = 1, \dots \tag{49}$$

The boundary value problem (43, 45, 47) was solved using the separation of variables to give

$$\theta\left(r,\varphi\right) = A_0 + B_0 \ln r + \sum_{n=1}^{\infty} \left(C_n r^n + D_n r^{-n}\right) \cos n\varphi. \tag{50}$$

Where

$$A_0 = \frac{q_0 r_o(\varphi)}{k}\left(\frac{1}{Bi} - \ln a\right),$$ (51)

$$B_0 = \frac{q_0 r_o(\varphi)}{k},$$ (52)

$$C_n = \frac{q_n r_o(\varphi)}{k}\;\frac{\frac{1}{n}u^n\left(Bi+n\right)\frac{1}{a^n}}{Bi\left(u^{2n}+1\right)+n\left(u^{2n}-1\right)},$$ (53)

$$D_n = -\frac{q_n r_o(\varphi)}{k}\;\frac{\frac{1}{n}u^n\left(Bi-n\right)a^n}{Bi\left(u^{2n}+1\right)+n\left(u^{2n}-1\right)}.$$ (54)

The ratio of the outer to inner radius of the eccentric flux tube: $u = u(\varphi) = r_o(\varphi)/a$ depends on the angle φ, since the outer radius of the tube flux

$$r_o = e\cos\varphi + \sqrt{b^2 - \left(e\sin\varphi\right)^2}$$ (55)

is the function of the angle φ.

Eq. (50) can be used for the temperature calculation when all the boundary conditions are known. In the inverse heat conduction problem three parameters are to be determined:

- absorbed heat flux referred to the projected furnace wall surface: $x_1 = q_m$,
- heat transfer coefficient on the inner surface of the boiler tube: $x_2 = h$,
- fluid bulk temperature: $x_3 = T_f$.

These parameters appear in boundary conditions (44) and (45) and will be determined based on the wall temperature measurements at m internal points (r_i, φ_i)

$$T\left(r_i, \varphi_i\right) = f_i \, , i = 1, ..., m \, , m \geq 3. \tag{56}$$

In a general case, the unknown parameters: $x_1, ..., x_n$ are determined by minimizing sum of squares

$$S = \left(\mathbf{f} - \mathbf{T_m}\right)^T \left(\mathbf{f} - \mathbf{T_m}\right), \tag{57}$$

where $\mathbf{f} = (f_1, ..., f_m)^T$ is the vector of measured temperatures, and $\mathbf{T_m} = (T_1, ..., T_m)^T$ the vector of computed temperatures $T_i = T(r_i, \varphi_i)$, $i = 1, ..., m$.

The parameters $x_1... x_n$, for which the sum (34) is minimum are determined using the Levenberg-Marquardt method [23,25]. The parameters, $\mathbf{x}$, are calculated by the following iteration

$$\mathbf{x}^{(k+1)} = \mathbf{x}^{(k)} + \boldsymbol{\delta}^{(k)}, \ k = 0,1,.... \tag{58}$$

Where

$$\boldsymbol{\delta}^{(k)} = \left[\left(\mathbf{J}_m^{(k)}\right)^T \mathbf{J}_m^{(k)} + \mu^{(k)}\mathbf{I}_n\right]^{-1} \times$$
$$\times \left(\mathbf{J}_m^{(k)}\right)^T \left[\mathbf{f} - \mathbf{T}_m\left(\mathbf{x}^{(k)}\right)\right]. \tag{59}$$

where $\mu(k)$ is the multiplier and $\mathbf{I}_n$ is the identity matrix. The Levenberg–Marquardt method is a combination of the Gauss–Newton method ($\mu^{(k)} \to 0$) and the steepest-descent method ($\mu^{(k)} \to \infty$). The $m \times n$ Jacobian matrix of $T(\mathbf{x}^{(k)}, \mathbf{r}_i)$ is given by

$$\mathbf{J}^{(k)} = \left.\frac{\partial \mathbf{T}(\mathbf{x})}{\partial \mathbf{x}_T}\right|_{\mathbf{x}=\mathbf{x}^{(k)}} = \left[\begin{array}{ccc} \dfrac{\partial T_1}{\partial x_1} & \cdots & \dfrac{\partial T_1}{\partial x_n} \\ \cdots & \cdots & \cdots \\ \cdots & \cdots & \cdots \\ \cdots & \cdots & \cdots \\ \cdots & \cdots & \cdots \\ \dfrac{\partial T_m}{\partial x_1} & \cdots & \dfrac{\partial T_m}{\partial x_n} \end{array}\right]_{\mathbf{x}=\mathbf{x}^{(k)}}, \quad m=5, \quad n=3, \tag{60}$$

The symbol $\mathbf{I}n$ denotes the identity matrix of $n \times n$ dimension, and $\mu^{(k)}$ the weight coefficient, which changes in accordance with the algorithm suggested by Levenberg and Marquardt. The upper index T denotes the transposed matrix. Temperature distribution $T(r, \varphi, \mathbf{x}^{(k)})$ is computed at each iteration step using Eq. (50). After a few iterations we obtain a convergent solution.

The Uncertainty of the Results

The uncertainties of the determined parameters $\mathbf{x}^*$ will be estimated using the error propagation rule of Gauss [23-26]. The propagation of uncertainty in the independent variables: measured wall temperatures f_j, $j=1, \ldots m$, thermal conductivity k, radial and angular positions of temperature sensors r_j, φj, $j=1, \ldots m$ is estimated from the following equation

$$2\sigma_{x_i} = \left[\sum_{j=1}^{m}\left(\frac{\partial x_i}{\partial f_j}\sigma_{f_j}\right)^2 + \left(\frac{\partial x_i}{\partial k}\sigma_k\right)^2 + \sum_{j=1}^{m}\left(\frac{\partial x_i}{\partial r_j}\sigma_{r_j}\right)^2 + \sum_{j=1}^{m}\left(\frac{\partial x_i}{\partial \varphi_j}\sigma_{\varphi_j}\right)^2\right]^{1/2}, \tag{61}$$

$$i = 1,2,3$$

The 95% uncertainty in the estimated parameters can be expressed in the form

$$x_i = x_i^* \pm 2\sigma_{x_i},$$

(62)

where $x_i^*, i = 1,2,3$ represent the value of the parameters obtained using the least squares method. The sensitivity coefficients $\partial x_i / \partial f_j,\ \partial x_i / \partial k,\ \partial x_i / \partial r_i,$ in the expression (61) were calculated by means of the numerical approximation using central difference quotients:

$$\frac{\partial x_i}{\partial f_j} = \frac{x_i\left(f_1, f_2, \ldots, f_j + \delta, \ldots, f_m\right) - x_i\left(f_1, f_2, \ldots, f_j - \delta, \ldots, f_m\right)}{2\delta},$$

(63)

where δ is a small number.

Computational and Boiler Tests

Firstly, a computational example will be presented. "Experimental data" are generated artificially using the analytical solution (50). Consider a water-wall tube with the following parameters (Figure1.):

- outer radius b = 35 mm,
- inner radius a = 25 mm,
- pitch of the water-wall tubes t = 80 mm,
- thermal conductivity k = 28.5 W/(m K),

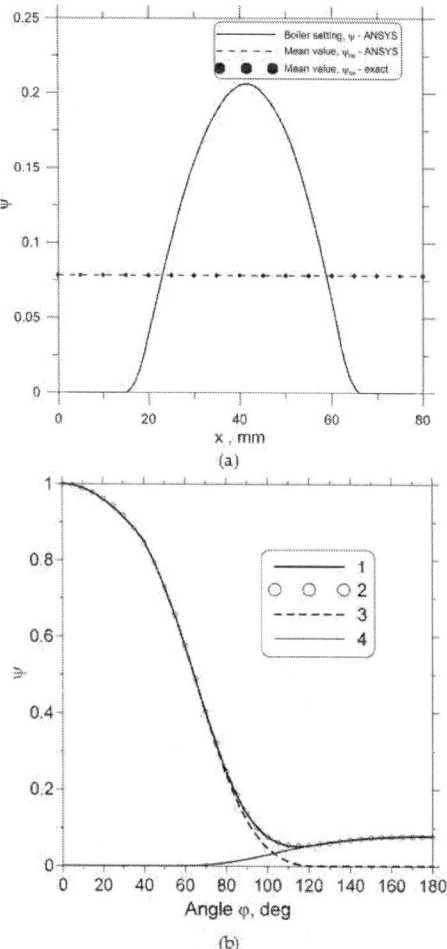

Figure 12: View factor associated with radiation heat exchange between elemental surface on the boiler setting or flux tube and flame: (a) – view factor for radiation heat transfer between flame and boiler setting, (b) 1 - total view factor accounting radiation from furnace and boiler setting, 2 - approximation by the Fourier polynomial of the seventh degree, 3 - exact view factor for furnace radiation, 4- view factor from boiler setting

- absorbed heat flux q_m = 200000 W/m²,
- heat transfer coefficient h = 30000 W/(m² K),
- fluid temperature T_f= 318 ºC.

The view factor distributions on the outer surface of the flux-tube and boiler setting were calculated analytically and numerically by means of the finite element method (FEM) [22]. The changes of the view factor over the pitch length and tube circumference are illustrated in Figures 12 and 13.

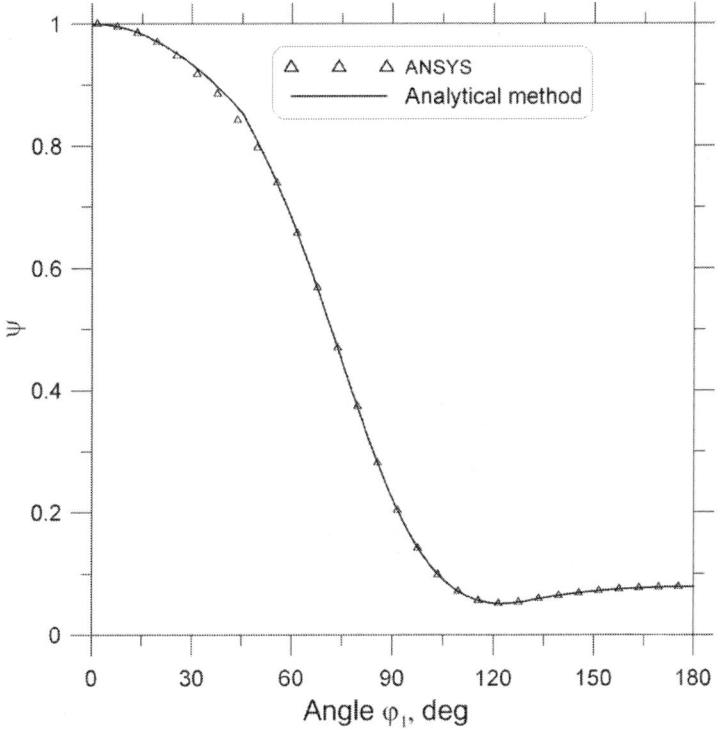

Figure 13: Comparison of total view factor calculated by exact and FEM method

The agreement between the temperatures of the outer and inner tube surfaces which were calculated analytically and numerically is also very good (Figures 14 and 15). The small differences between the analytical and FEM solutions are caused by the approximate boundary condition (47). The temperature distribution in the flux tube cross section is shown in Figure 14.

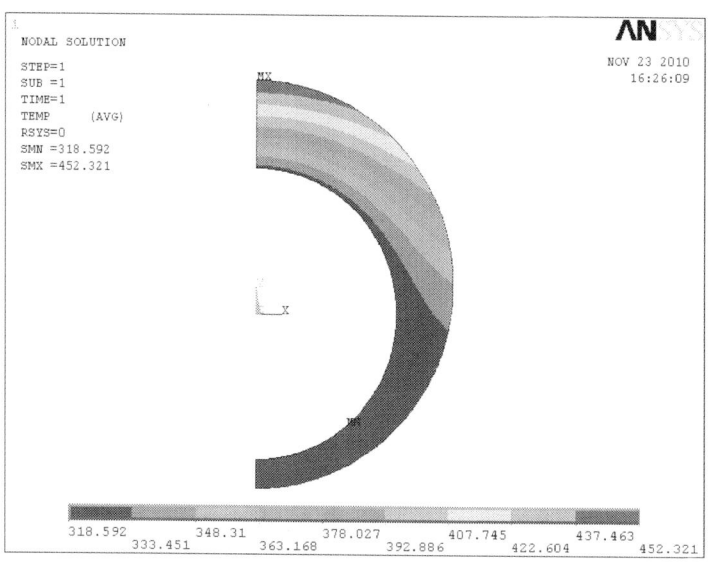

Figure 14: Computed temperature distribution in °C in the cross section of the heat flux tube; qm = 200000 W/m², h = 30000 W/(m²·K), Tf = 318 °C.

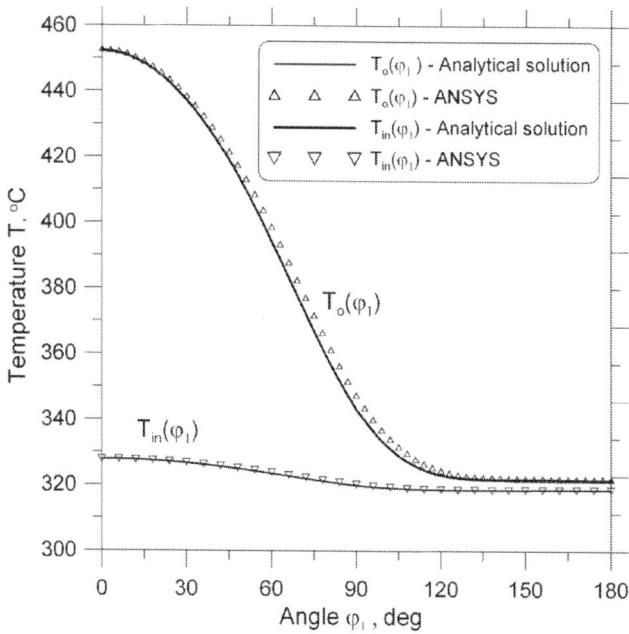

Figure 15: Temperature distribution at the inner and outer surfaces of the flux tube calculated by the analytical and finite element method.

The following input data is generated using Eq. (50):

$$f_1 = 437.98 \ ^\circ C, f_2 = 434.47 \ ^\circ C,$$
$$f_3 = 383.35 \ ^\circ C, f_4 = 380.70 \ ^\circ C, f_5 = 321.58 ^\circ C.$$

The following values were obtained using the proposed method:

$$\dot{q}_m = 200\,000.35 \ \text{W/m}^2, h^* = 30\,001.56 \ \text{W/(m}^2 \cdot \text{K)}, T_f^* = 318.00 \ ^\circ C.$$

In order to show the influence of the measurement errors on the determined thermal boundary parameters, the 95% confidence intervals were calculated. The following uncertainties of the measured values were assumed (at a 95% confidence interval):

$$2\sigma_{f_j} = \pm 0.2K, j = 1, \dots, 5, 2\sigma_k = \pm 0.5 \ \text{W/(m·K)}, 2\sigma_{r_i} = \pm 0.05 \text{mm}, 2\sigma_{\varphi_j} = \pm 0.5^\circ, j = 1, \dots, 5.$$

The uncertainties (95% confidence interval) of the coefficients x_i were determined using the error propagation rule formulated by Gauss.

The calculation using Eq. (61) yielded the following results: $x_1 = 200\,000.35 \pm 3827.72 \ \text{W/m}^2$, $x_2 = 30\,001.56 \pm 2698.81 \ \text{W/(m}^2 \ \text{K)}$, $x_3 = 318.0 \pm 0.11 \ ^\circ C$. The accuracy of the obtained results is very satisfactory. There is only a small difference between the estimated parameters and the input values. The highest temperature occurs at the crown of the flux-tube (Figures 14 and 15). The temperature of the inner surface of the flux tube is only a few degrees above the saturation temperature of the water-steam mixture. Since the heat flux at the rear side of the tube is small, the circumferential heat flow rate is significant. However, the rear surface thermocouple indicates temperatures of 2-4 °C above the saturation temperature. Therefore, the fifth thermocouple can be attached to the unheated side of the tube so as to measure the temperature of the water-steam mixture flowing through the flux tube.

In the second example, experimental results will be presented. Measurements were conducted at a 50MW pulverized coal fired boiler. The temperatures indicated by the flux tube at the elevation of 19.2 m are shown in Figure 16. The heat flux tube is of 20G low carbon steel with temperature dependent thermal conductivity

$$k(T) = 53.26 - 0.02376224T, \tag{64}$$

where the temperature T is expressed in ºC and thermal conductivity in W/(m K).

The unknown parameters were determined for eight time points which are marked in Figure16.

The inverse analysis was performed assuming the constant thermal conductivity $k(\bar{T})$ which was obtained from Eq. (64) for the average temperature: $\bar{T} = (T_1 + T_2 + T_3 + T_4)/4$.

The estimated parameters: heat flux q_m, heat transfer coefficient h, and the water-steam mixture Tf are depicted in Figure17.The developed flux tube can work for a long time in the destructive high temperature atmosphere of a coal-fired boiler.

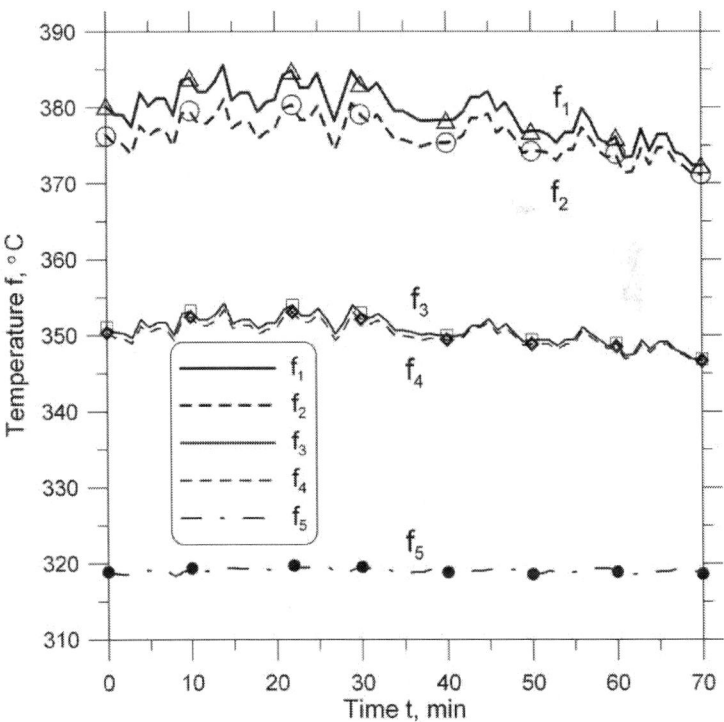

Figure 16: Measured flux tube temperatures; marks denote measured temperatures taken for the inverse analysis.

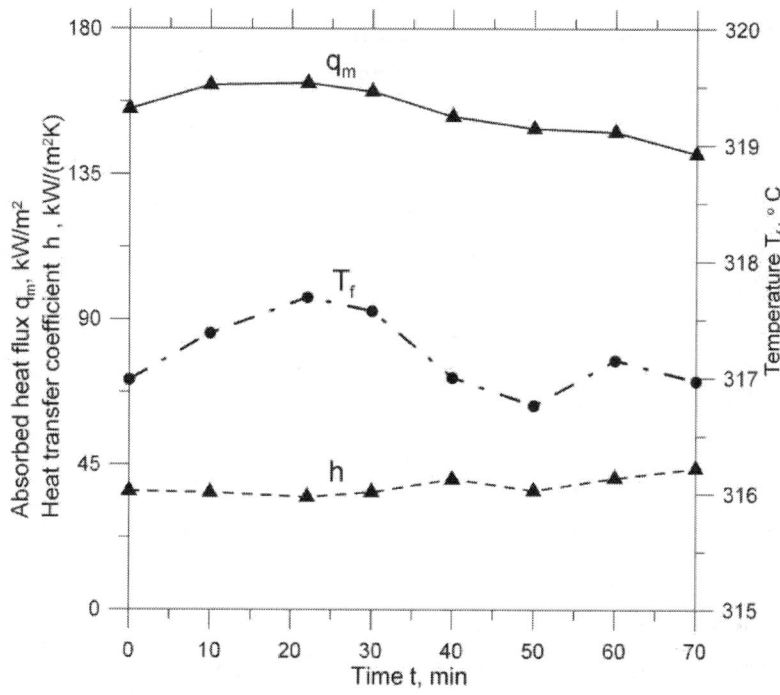

Figure 17: Estimated parameters: absorbed heat flux qm, heat transfer coefficient h, and temperature of water-steam mixture Tf

Flux tubes can also be used as a local slag monitor to detect a build up of slag. The presence of the scale on the inner surface of the tube wall can also be detected.

TUBULAR TYPE HEAT FLUX METER MADE OF A FINNED TUBE

In this section, a numerical method for determining the heat flux in boiler furnaces, based on experimentally acquired interior flux-tube temperatures, is presented. The tubular type instrument has been designed (Figure 18) to provide a very accurate measurement of absorbed heat flux q_m, inside heat transfer coefficient h_{in}, and water steam temperature T_f. The number of thermocouples is greater than three because the additional information can help enhance the accuracy of parameter determining. In contrast to the existing devices, in the developed flux-tube fins are not welded to adjacent water-wall tubes. Temperature distribution in the flux-tube is symmetric and not disturbed by different temperature fields in neighboring tubes. The temperature dependent thermal conductivity of the flux-tube material was assumed. The meter is constructed from a short length of eccentric tube containing four thermocouples on the fireside below the inner and outer surfaces of the tube. The fifth thermocouple is located at the rear of the tube (on the casing side of the water-wall tube). The boundary conditions on the outer and inner surfaces of the water flux-tube must then be determined from temperature measurements in the interior locations. Four K-type sheathed thermocouples, 1 mm in diameter, are inserted into holes, which are parallel to the tube axis. The thermal conduction effect at the hot junction is minimized because the thermocouples pass through isothermal holes. The thermocouples are brought to the rear of the tube in the slot machined in the tube wall. An austenitic cover plate with the thickness of 3 mm – welded to the tube – is used to protect the thermocouples from the incident flame radiation. A K-type sheathed thermocouple with a pad is used to measure the temperature at the rear of the flux-tube. This temperature is almost the same as the water-steam temperature. An inverse problem of heat conduction was solved using the least squares method. Three unknown parameters were estimated using the Levenberg-Marquardt method [23, 25]. At every iteration step, the temperature distribution over the cross-section of the heat flux meter was computed using the ANSYS CFX software

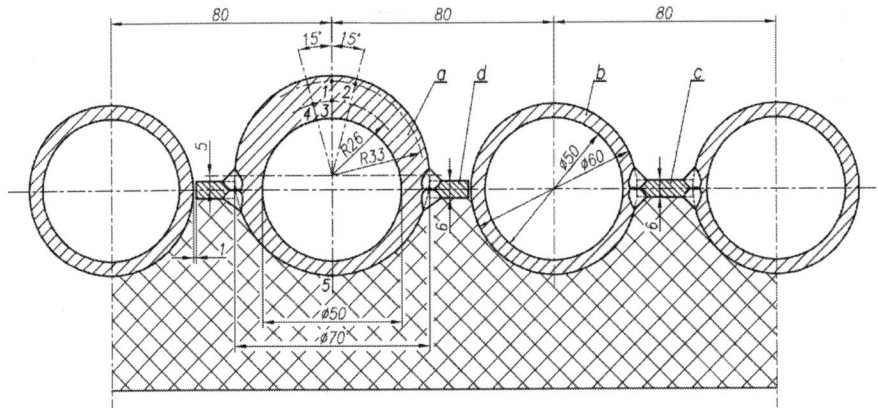

Figure 18: The cross-section of the membrane wall in the combustion chamber of the steam boiler

Test calculations were carried out to assess accuracy of the presented method. The uncertainty in determined parameters was calculated using the Gauss variance propagation rule. The presented method is appropriate for membrane water walls (Figure 18). The new method has advantages in terms of simplicity and flexibility.

Theory

The furnace wall tubes in most modern units are welded together with steel bars (fins) to provide membrane wall panels which are insulated on one side and exposed to a furnace on the other, as shown schematically in Figure 18.

In a heat conduction model of the flux-tube the following assumptions are made:

- temperature distribution is two-dimensional and steady-state,
- the thermal conductivity of the flux-tube and membrane wall,
- may be dependent of temperature,
- the heat transfer coefficient *hin* and the scale thickness ds is uniform over the inner tube surface.

The temperature distribution is governed by the non-linear partial differential equation

$$\nabla \cdot \left[k(T) \nabla T \right] = 0, \tag{65}$$

where ∇ is the vector operator, which is called nabla (gradient operator), and in Cartesian coordinates is defined by $\nabla = \mathbf{i} \partial / \partial x + \mathbf{j} \partial / \partial y + \mathbf{k} \partial / \partial z +$. The unknown boundary conditions may be expressed as

$$\left[k(T) \frac{\partial T}{\partial n} \right]_s = q(s), \tag{66}$$

where $q(s)$ is the radiation heat flux absorbed by the exposed fluxtube and membrane wall surface. The local heat flux $q(s)$ is a function of the view factor $\psi(s)$ (Figure 19)

$$q(s) = q_m \psi(s), \tag{67}$$

where qm is measured heat flux (thermal loading of heating surface). The view factor $\psi(s)$ from the infinite flame plane to the differential element on the membrane wall surface can be determined graphically[7], or numerically [22].

In this chapter, $\psi(s)$ was evaluated numerically using the finite element program ANSYS [22], and is displayed in Figure 19 as a function of the extended coordinate s. Because of the symmetry, only the representative water-wall section illustrated in Figure 20 needs to be analyzed. The convective heat transfer from the inside tube surfaces to the water-steam mixture is described by Newton's law of cooling

$$-\left[k(T) \frac{\partial T}{\partial n} \right]_{s_{in}} = h_{in} \left(T \big|_{s_{in}} - T_f \right), \tag{68}$$

where $\partial T/\partial n$ is the derivative in the normal direction, h_{in} is the heat transfer coefficient and T_f denotes the temperature of the water–steam mixture.

The reverse side of the membrane water-wall is thermally insulated. In addition to the unknown boundary conditions, the internal temperature measurements f_i are included in the analysis

$$T_e\left(\mathbf{r}_i\right) = f_i, \quad i = 1,\dots,m, \tag{69}$$

where $m = 5$ denotes the number of thermocouples (Figure 18). The unknown parameters: $x_1 = q_m$, $x_2 = h_{in}$, and $x_3 = T_f$ were determined using the least-squares method. The symbol r_{in} denotes the inside tube radius, and $k(T)$ is the temperature dependent thermal conductivity. The object is to choose $\mathbf{x} = (x_1, \dots, x_n)^T$ for $n = 3$ such that computed temperatures $T(\mathbf{x}, \mathbf{r}_i)$ agree within certain limits with the experimentally measured temperatures f_i.

This may be expressed as

$$T\left(\mathbf{x},\mathbf{r}_i\right) - f_i \cong 0, \quad i = 1,\dots,m, \quad m = 5. \tag{70}$$

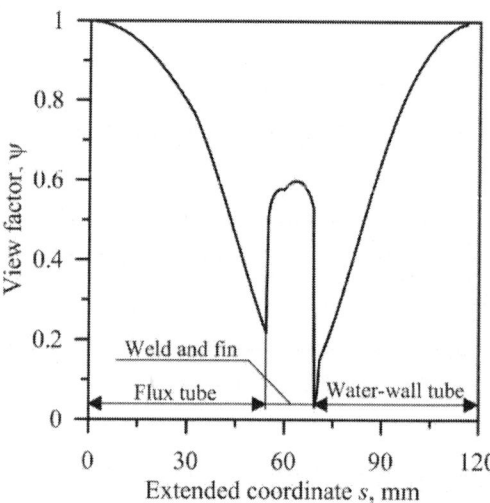

Figure 19: View factor distribution on the outer surface of water-wall tube.

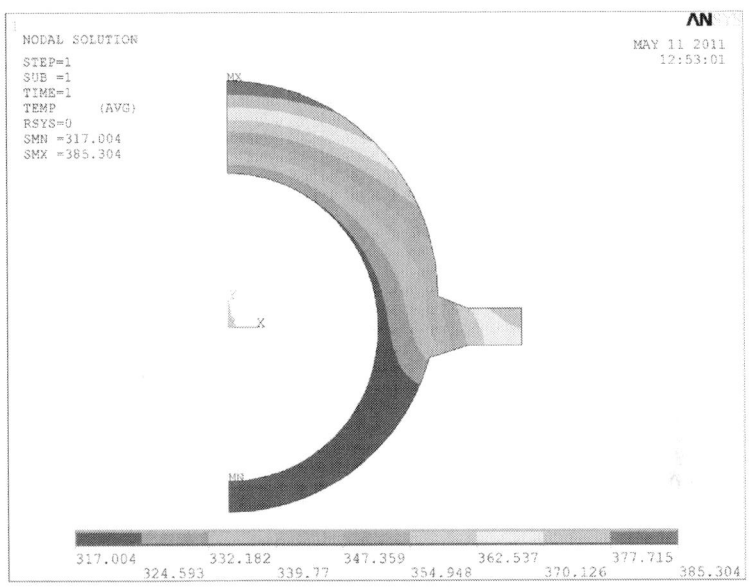

Figure 20: Temperature distribution in the flux tube cross-section for: qm = 150000 W/m², hin = 27000 W/(m²K) and T_f = 317°C

The least-squares method is used to determine parameters **x**. The sum of squares

$$S = \sum_{i=1}^{m} \left[f_i - T(\mathbf{x}, \mathbf{r}_i) \right]^2, \quad m = 5, \tag{71}$$

is minimized usingthe Levenberg–Marquardt method [23, 25].

The uncertainties of the determined parameters **x*** will be estimated using the error propagation rule of Gauss [23-26].

Test Computations

The flux-tubes were manufactured in the laboratory and then securely welded to the water-wall tubes at different elevations in the furnace of the steam boiler. The coal fired boiler produces 58.3 kg/s superheated steam at 11 MPa and 540°C.

The material of the heat flux-tube is 20G steel. The composition of the 20G mild steel is as follows: 0.17–0.24% C, 0.7–1.0% Mn, 0.15–0.40% Si, Max 0.04% P, Max 0.04% S, and the remainder is iron Fe. The heat flux-tube thermal conductivity is assumed to be temperature dependent (Table 1).

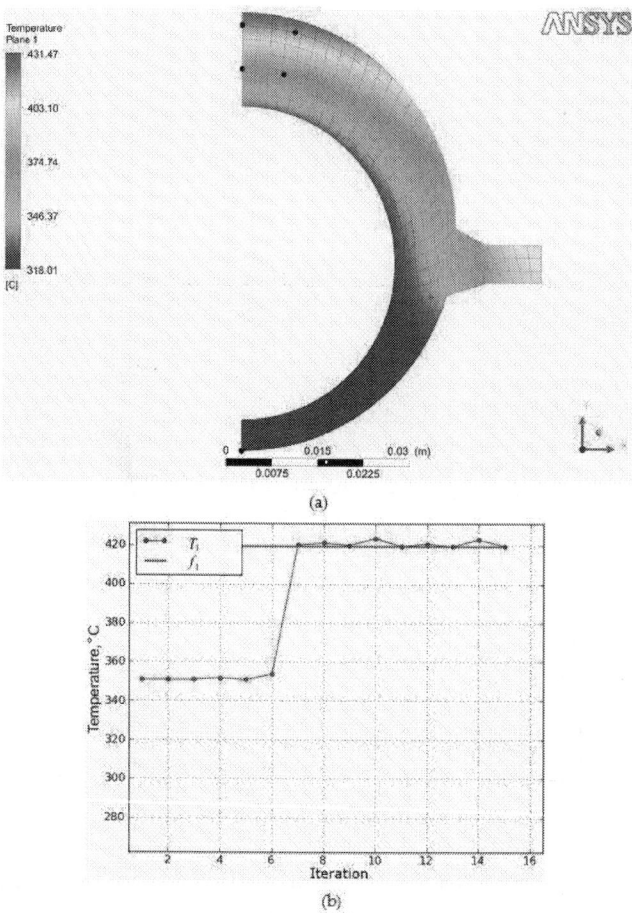

Figure 21: Solution of the inverse problem for the "exact" data: f_1 = 419.66°C, f_2 = 417.31°C, f_3 = 374.90°C, f_4= 373.19°C, f_5 = 318.01 °C ; (a) - temperature distribution in the flux-tube, (b) - iteration number for the temperature T_1

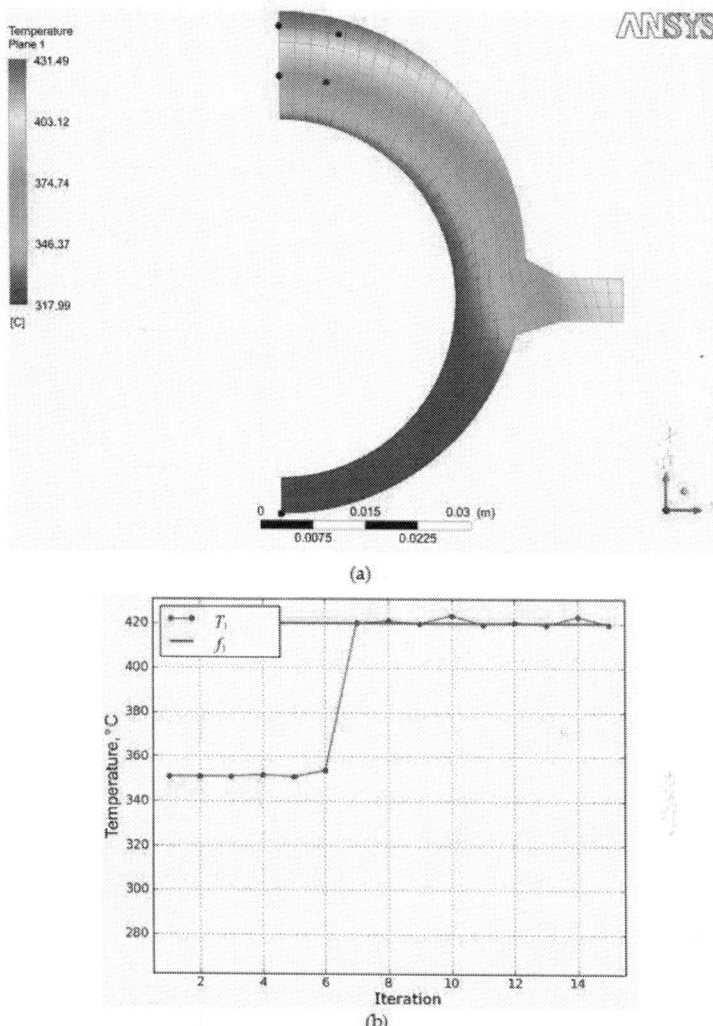

Figure 22: Solution of the inverse problem for the "perturbed" data: f_1 = 420.16°C, f_2 = 416.81°C, f_3 = 375.40°C, f_4 = 372.69°C, f_5 = 318.01°C;(a) - temperature distribution in the flux-tube, (b) - iteration number for the temperature T_1

Table 1: Thermal conductivity $k(T)$ of steel 20G as a function of temperature

Temperature T, °C	100	200	300	400
Thermal conductivity k, W/(m·K)	50.69	48.6	46.09	42.3

To demonstrate that the maximum temperature of the fin tip is lower than the allowable temperature for the 20G steel, the flux tube temperature was computed using ANSYS CFX package [22]. Changes of the view factor on the flux tube, weld and fin surface were calculated with ANSYS CFX. The temperature distribution shown in Figure 20 was obtained for the following data: absorbed heat flux, q_m = 150000 W/m², temperature of the water-steam mixture, T_f = 317°C, and heat transfer coefficient at the tube inner surface, h_{in} = 27000 W/(m² K). An inspection of the results shown in Figure 20indicates that the maximum temperature of the fin does not exceed 375°C.

Next, to illustrate the effectiveness of the presented method, test calculations were carried out. The "measured" temperatures f_i, i = 1, 2, ..., 5 were generated artificially by means of ANSYS CFX for: q_m = 250000 W/m², h_{in} = 30000 W/(m² K) and T_f = 318°C. The following values of "measured" temperatures were obtained f_1 = 419.66°C, f_2 = 417.31°C, f_3 = 374.90°C, f_4 = 373.19°C, f_5 = 318.01°C. The temperature distribution in the flux tube cross-section, reconstructed on the basis of five measured temperatures is depicted in Figure 21a.

The proposed inverse method is very accurate since the estimated parameters: q_m = 250000.063 W/m², h_{in} = 30000.054 W/(m² K)and Tf = 318.0°C differ insignificantly from the input values. In order to show the influence of the measurement errors on the determined parameters, the 95% confidence intervals were estimated. The following uncertainties of the measured values were assumed (at 95% confidence interval): $2\sigma_{f_j} = \pm 0.5 \text{ K}, \; j = 1, 2, ..., 5,$ $2\sigma_k = \pm 1 \text{ W/(m·K)}, 2\sigma_{r_i} = \pm 0.05\text{mm}, \; 2\sigma_{\varphi_i} = \pm 0.5°, \; j=1,...,5.$

The uncertainties (95% confidence interval) of the coefficients x_i were determined using the error propagation rule formulated by Gauss [23-26]. The calculated uncertainties are: ±6% for q_m, ±33% for h_{in} and ±0.3% for T_f. The accuracy of the results obtained is acceptable.

Then, the inverse analysis was carried out for perturbed data: f_1 = 420.16°C, f_2 = 416.81°C, f_3 = 375.40°C, f_4 = 372.69°C, f_5 = 318.01°C. The reconstructed temperature distribution illustrates Figure 22a.

The obtained results are: q_m = 250118.613 W/m², h_{in} = 30050.041 W/(m² K) and Tf = 317.99°C. The errors in the measured temperatures have little effect on the estimated parameters. The number of iterations in the Levenberg-Marquardt procedure is small in both cases (Figures 21b and 22b).

CONCLUSIONS

Two different tubular type instruments (flux tubes) were developed to identify boundary conditions in water wall tubes of steam boilers. The first measuring device is an eccentric tube. The ends of the four thermocouples are located at the fireside part of the tube and the fifth thermocouple is attached to the unheated rear surface of the tube. The meter presented in the paper has one particular advantage over the existing flux tubes to date. The temperature distribution in the flux tube is not affected by the water wall tubes, since the flux tube is not connected to adjacent waterwall tubes with metal bars, referred to as membrane or webs. To determine the unknown parameters only the temperature distribution at the cross section of the flux tube must be analyzed.

The second flux tube has two longitudinal fins. Fins attached to the flux tube are not welded to the adjacent water-wall tubes, so the temperature distribution in the measuring device is not affected by neighboring water-wall tubes. The installation of the flux tube is easier because welding of fins to adjacent water-wall tubes is avoided. Based on the measured flux tube temperatures the non-linear inverse heat conduction problem was solved. A CFD based method for determining heat flux absorbed water wall tubes, heat transfer coefficient at the inner flux tube surface and temperature of the water-steam mixture has been presented. The proposed flux tube and the inverse procedure for determining absorbed heat flux can be used both when the inner surface of the heat flux tube is clean and when scale or corrosion deposits are present on the inner surface what can occur after a long time service of the heat flux tube.

REFERENCESS

1. M. Segeer, J. Taler, 1983Konstruktion und Einsatz transportabler Wärmeflußsonden zur Bestimmung der Heizflächenbelastung in Feuerräumen. Fortschr.-Ber. VDI Zeitschrift, Reihe 6, Nr 129. Düsseldorf : VDI-Verlag.
2. Northover EW, Hitchcock JA1967Measurements of Local Heat Flux and Water-Side Heat Transfer Coefficient in Water Wall TubesJ. Sci. Instrum. 44371374
3. Neal SBH, Northover EW1980The Measurement of Radiant Heat Flux in Large Boiler Furnaces-I. Problems of Ash Deposition Relating to Heat Flux. Int. J. Heat Mass Transfer 2310151022
4. N. Arai, A. Matsunami, S. Churchill, 1996Measurements of Local Heat Flux and Water-Side Heat Transfer Coefficient in Water Wall TubesExp. Therm. Fluid Sci. 12452460
5. J. Taler, 1990Measurement of Heat Flux to Steam Boiler Membrane Water Walls. VGB Kraftwerkstechnik 70540546
6. J. Taler, 1992Measurements of Local Heat Flux and Water-Side Heat Transfer Coefficient in Water Wall TubesInt. J. Heat Mass Transfer 3516251634

7. J. Taler, 1990Messung der lokalen Heizflächenbelastung in Feuerräumen von Dampferzeugern. Brennstoff-Wärme-Kraft (BWK) 42269277

8. Z. Fang, D. Xie, N. Diao, J. R. Grace, C. J. Lim, 1997Measurements of Local Heat Flux and Water-Side Heat Transfer Coefficient in Water Wall TubesInt. J. Heat Mass Transfer 4039473953

9. W. Luan, Lim. C. J. Bowen, C. M. H. Brereton, J. R. Grace, 2000Suspension-to Membrane-Wall Heat Transfer in a Circulating Fluidized Bed Combustor. Int. J. Heat Mass Transfer 4311731185

10. J. Taler, D. Taler, 2007Measurements of Local Heat Flux and Water-Side Heat Transfer Coefficient in Water Wall TubesHeat Transfer Engineering28230239

11. T. Sobota, D. Taler, 2010Measurements of Local Heat Flux and Water-Side Heat Transfer Coefficient in Water Wall TubesRynek Energii86108114

12. D. Taler, J. Taler, A. Sury, 2011Identification of Thermal Operation Conditions of Water Wall Tubes Using Eccentric Tubular Type Meters. Rynek Energii 92164171

13. J. Taler, D. Taler, A. Kowal, 2011Measurements of Local Heat Flux and Water-Side Heat Transfer Coefficient in Water Wall TubesArchives of Thermodynamics327788

14. J. Taler, D. Taler, T. Sobota, P. Dzierwa, 2011Measurements of Local Heat Flux and Water-Side Heat Transfer Coefficient in Water Wall TubesArchives of Thermodynamics32103116

15. LeVert FE, Robinson JC, Frank RL, Moss RD, Nobles WC, Anderson AA1987A Slag Deposition Monitor for Use in Coal_Fired Boilers. ISA Transactions 265164

16. LeVert FE, Robinson JC, Barrett SA, Frank RL, Moss RD, Nobles WC, Anderson AA1988Slag Deposition Monitor for Boiler Performance Enhancement. ISA Transactions 275157

17. A. Vallero, C. Cortes, 1996Ash Fouling in Coal-Fired Utility Boilers. Monitoring and Optimization of On-Load Cleaning. Prog. Energy. Combust. Sci. 22189200

18. E. Teruel, C. Cortes, L. I. Diez, I. Arauzo, 2005Measurements of Local Heat Flux and Water-Side Heat Transfer Coefficient in Water Wall TubesChem. Eng. Sci. 6050355048

19. J. Taler, M. Trojan, D. Taler, 2011Monitoring of Ash Fouling and Internal Scale Deposits in Pulverized Coal Fired Boilers. New York: Nova Science Publishers.
20. J. R. Howell, R. Siegel, M. P. Mengüç, 2011Measurements of Local Heat Flux and Water-Side Heat Transfer Coefficient in Water Wall TubesBoca Raton: CRC Press- Taylor & Francis Group.
21. Sparrow FM, Cess RD1978Measurements of Local Heat Flux and Water-Side Heat Transfer Coefficient in Water Wall TubesNew York: McGraw-Hill.
22. ANSYS CFX 12.2010Urbana, Illinois, USA: ANSYS Inc.
23. Seber GAF,Wild CJ1989Measurements of Local Heat Flux and Water-Side Heat Transfer Coefficient in Water Wall TubesNew York: Wiley.
24. Policy on reporting uncertainties in experimental measurements and results2000ASME J. Heat Transfer 122411413
25. Press WH, Teukolsky SA, Vetterling WT, Flannery BP2006Measurements of Local Heat Flux and Water-Side Heat Transfer Coefficient in Water Wall TubesCambridge: Cambridge University Press.
26. Coleman HW, Steele WG2009Measurements of Local Heat Flux and Water-Side Heat Transfer Coefficient in Water Wall Tubesfor Engineers. Hoboken: Wiley.

CITATION

Jan Taler and Dawid Taler (2012). Measurements of Local Heat Flux and Water-Side Heat Transfer Coefficient in Water Wall Tubes, An Overview of Heat Transfer Phenomena, Dr M. Salim Newaz Kazi (Ed.), ISBN: 978-

CHAPTER 3

Numerical Study of Heat Transfer and Contaminant Transport in an Unsaturated Porous Soil

Abdelhamid Belghit[1]*, Mustapha Benyaich[2]

[1]Laboratory of Engineering Science for Environment—LaSIE FRE-CNRS 3474, University of La Rochelle, La Rochelle, France

[2]Department of Physics, Faculty of Science, University Cadi Ayyad, Marrakesh, Morocco

ABSTRACT

Penetration of chemicals in the soil ground through irrigation water or rainfall induces important risks for the environment. These risks are badly known and may lead to direct contamination of the environment (atmosphere or ground water) or harmful effects on organisms living at ground level, indirectly affecting men. It is thus necessary to estimate these potential chemical risks on the environment. For that reason, the gradual change of these products (fertilizers, solutions, pollutants, ...) in the ground has been the subject of a lot of recent research works, based in particular on the study of non-saturated porous media in a theoretical, numerical or experimental way. Most of these works are incomplete and, in order to simplify the problem, they don't take into accounts some process, which may be of prime importance under particular natural conditions. Complexity of such studies results from their multidisciplinary nature. In this communication, we study simultaneous transport of pollutant, the water that provides transport and the heat transfer in a 200 cm long cylindrical column full of sand taken as a non-saturated porous medium. We consider two kinds of conditions on the temperature at the column surface: the case of constant temperature and the case of sinusoidal temperature. We evaluate the influence of this temperature on the transfers. This study is purely numerical. We use the control volume method to determine hydrous, thermal and pollutant concentration profiles.

Keywords
Soil, Heat and Mass Transfer, Porous Media, Contaminant

INTRODUCTION

The transfers in solutions in the ground result from a set of phenomena whose knowledge should make it possible to improve management of the irrigation and spreading in agricultural medium, double point of view of the prevention of the contamination of underground water and economy of fertilisers.

Among these phenomena, the purely physical aspect of the transfers constitutes a whole of convection and dispersion mechanisms whose study, in natural environment, is essential for the comprehension of pollutant evolution in the ground.

During these last decades, the study of the vertical and horizontal isothermal transfers of a contaminant in an unsaturated homogeneous porous medium has been the object of an important literature treating the numerical [1] - [6] and experimental [4] [7] - [10] aspect.

However, the majority of these authors considered that the temperature on the surface of the ground is constant [4]. In fact, the temperature of the ground presents continual variations under the influence of the climatic conditions, which are carried out in a relatively stable way according to periodic cycles. These cycles, daily and annual, are indeed connected with the variations of intensity of the solar radiation which appear over some daily periods, by the alternation of the day and the night, and over annual periods, by the evolution of the average slope of the sun on the horizon. The sinusoidal approximation constitutes the simplest manner to describe mathematically a periodic behavior.

So we are going to solve numerically the coupled and non-linear equations governing these transfers, by the method of control volume worked out by Patankar [11].

The results will be given in the form of curves representing the evolution of the hydrous, thermal and concentration profiles. The object of these results is to contribute to the prediction and the control of the pollution of the ground water.

In this work, we will use two different boundary conditions of temperature. The first one is constant and equal to 40°C; the other one is sinusoidal given by real climatic measurements of Marrakesh (Morocco).

MATHEMATICAL MODEL

The equations of transfer of water, heat and concentration are given respectively by the model of Philip and Devries [12] and the equation of convection-dispersion, which under certain conditions [13], can be written under the following form:

$$\frac{\partial \theta}{\partial t} = \frac{\partial}{\partial z}\left[D_\theta \frac{\partial \theta}{\partial z} + D_T \frac{\partial T}{\partial z} - K(\theta)\right] \tag{1}$$

$$\rho C_p \frac{\partial T}{\partial t} = \frac{\partial}{\partial z}\left[\lambda \frac{\partial T}{\partial z}\right] \tag{2}$$

$$\theta \frac{\partial C}{\partial t} = \frac{\partial}{\partial z}\left[D_{ap}\theta \frac{\partial C}{\partial z}\right] - q_\theta \frac{\partial C}{\partial z} \tag{3}$$

The thermohydric coefficients used were evaluated using the experimental data collected by Crausse [14].

Initial and Boundary Conditions

To solve these equations it is necessary to know the initial and boundary conditions, we propose the following Dirichlet type conditions (Figure 1).

a) For constant surface temperature

$$\text{At } t = 0 \begin{cases} T = T_0 \\ \theta = \theta_0 \\ C = 0 \end{cases}$$;

$$\text{at } Z = 0 \text{ and } t > 0 \begin{cases} T = T_1 \\ \theta = \theta_s \\ C = C_0 \end{cases}$$

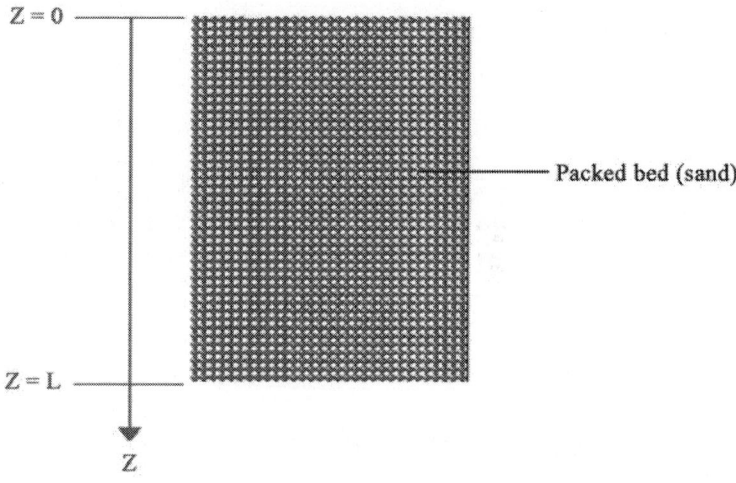

Packed bed (sand)

Figure 1: The column scheme.

$$\text{at } Z = L \text{ and } t > 0 \begin{cases} -\lambda \dfrac{\partial T}{\partial z} = 0 \\ -D_\theta \dfrac{\partial \theta}{\partial z} - D_T \dfrac{\partial T}{\partial z} + K = 0 \\ \dfrac{\partial C}{\partial z} = 0 \end{cases}$$

b) For variable surface temperature
For the resolution of the equation of heat, the values of the temperature on the surface are given by Ouardi et al. [15]

$$T(z,t) = T_0 + \theta_0 e^{-kz} \cos(\omega t - kz)$$

Who used the annual climatic data of Marrakech of the time temperature, to choose 10 days known as standard days, as seen in Figure 2.

c) Equation of moisture
The column is subjected to a constant flow in take making it possible to carry it to a saturation moisture. The other end is maintained at initial moisture:

$$\text{At time}: t = 0 \quad \theta = \theta_0$$
$$\text{At } Z = 0 \quad \theta = \theta_s$$
$$\text{At } Z = L \quad \theta = \theta_0$$

The contribution of aqueous solutions can be continuous, the case of a table cloth polluted on the surface of the ground, as it can be in crenel, precipitation or irrigation with use of a fertilizer, by considering the time of injection of flow of infinite or finished aqueous solutions t_m:

$$0 < t < t_m \quad C(X,z,t) = C_0$$
$$t > t_m \quad C(X,z,t) = 0$$

d) Equation of the concentration
For the equation of concentration we suppose that the surface is maintained at a concentration in solution C_0, whereas the field is initially deprived of any aqueous solution:

$$\text{At time}: t = 0 \quad C = 0$$
$$\text{At } Z = 0 \quad C = C_0$$
$$\text{At } Z = L \quad C = 0$$

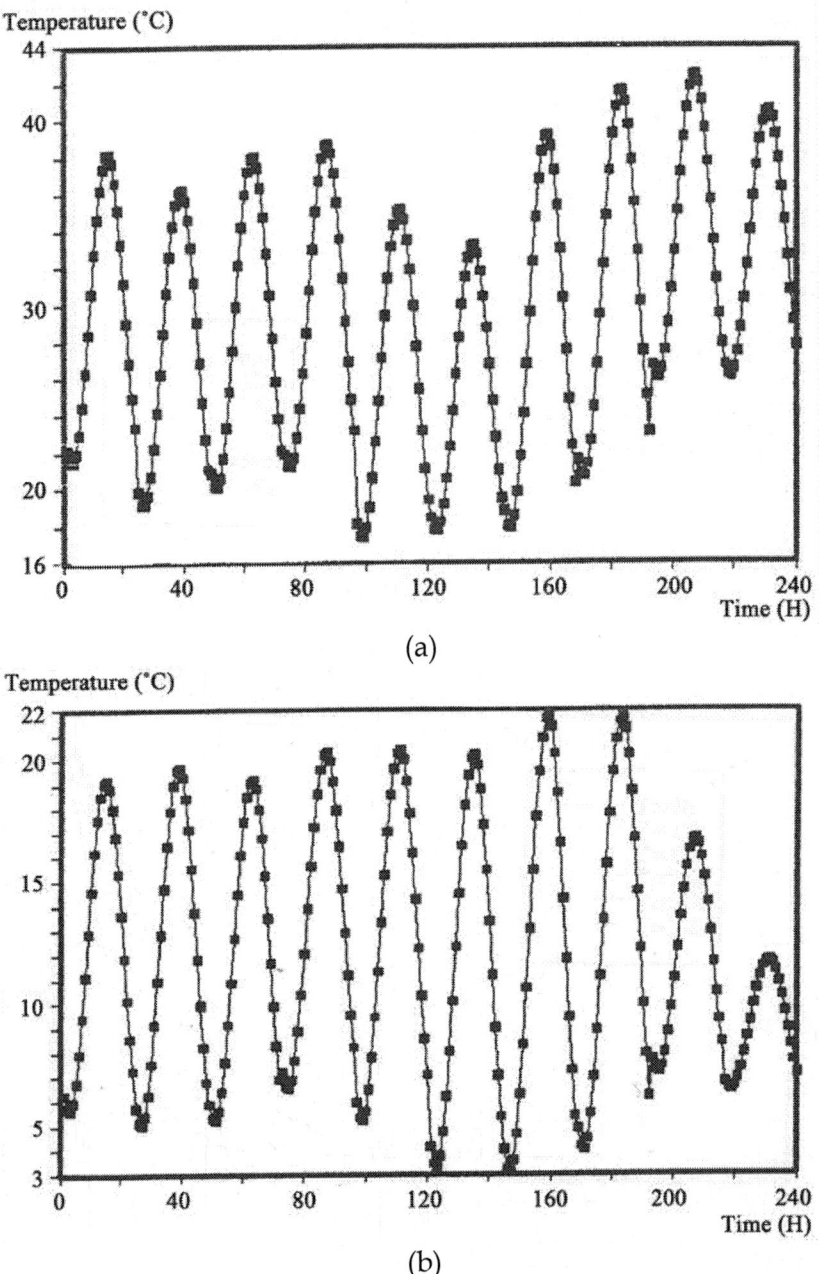

Figure 2: Values of the surface temperature taken from the annual climatic data of Marrakesh (Morocco) of the time given by Ouardi et al. [15] , (a) Summer; (b) Winter.

NUMERICAL SIMULATION

The numerical method used is that of the control volume proposed by Patankar [11] . It consists of the discretisation of the field by juxtaposition of elementary volumes inside which we integrates the equations to have them directly discretised. The temperature is periodic of period 24 hours. To show its influence on the transfers it is necessary to make simulations at least equal to this period. In this work, the time of simulation is 36 hours.

The parameters used are:

Δz = 0.5 cm step of space; Δt = 6 s, step of time; θ_0= 0.06 cm³/cm³; θ_s = 0.35 cm³/cm³; T_0 = 293 K; T_1 = 313 K; C_0 = 10 mg/l concentration of solute at the entry of column;

$$D_{ap} = \alpha\left(q_\theta / \theta\right) + a \cdot e^{b\theta}$$

And

$$\alpha = 0.2 \cdot L \cdot \left(1 - e^{(-0.05 Z/L)}\right),$$

Where, a and b are constants that depend on the nature of the soil [16].

RESULTS

In Figure 3 is presented the humidity variations with time and depth for a constant surface temperature. It takes 50 min to change the humidity at 40 cm of depth. The velocity of humidity penetration decreases with time due to the soil saturation.

Figure 4 shows the temperature evolution with time and depth, the profiles are similar to the humidity, but the velocity is much lower, this difference is principally to the small soil thermal conduction compared to the humidity dispersion.

The concentration of the inert pollutant is shown in Figure 5. For layers nearer to the soil surface the pollutant diffusion is fast, but for bigger depth the concentration changes are slower, and it changes its concentration only at 30 cm after 50 min.

Figure 6 shows the variation of temperature at various depths for one day in summer and winter. It is noticed that all the curves have a sinusoidal form but they present a dephasing compared to that of soil surface. With the increase in depth, the layers are less influenced by the heat transfer, thus their time to increase/decrease the temperature is bigger than in the soil surface. These results are in good agreement with those found by [3]

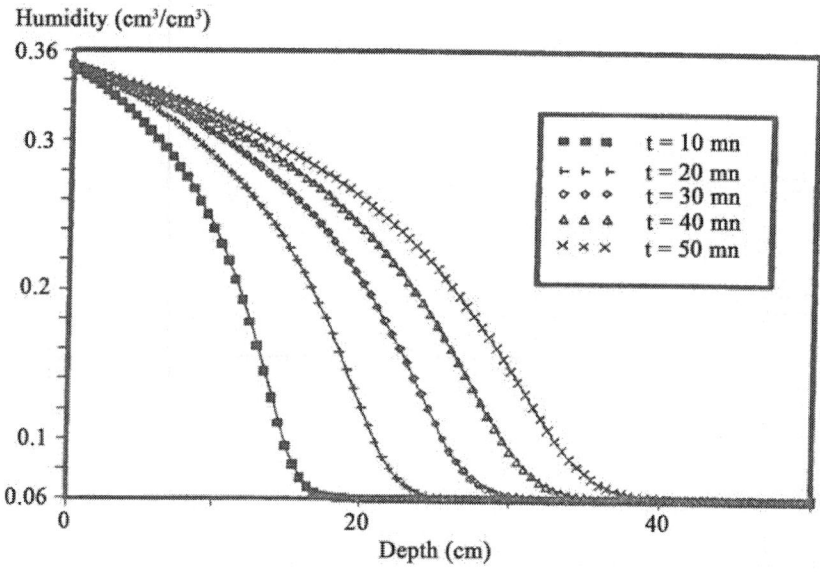

Figure 3: Humidity profiles with depth and time for a constant surface temperature.

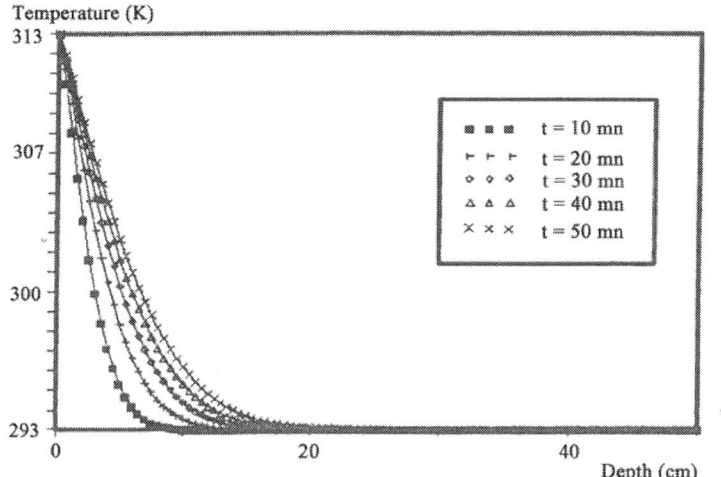

Figure 4: Temperature profiles with depth and time for a constant surface temperature.

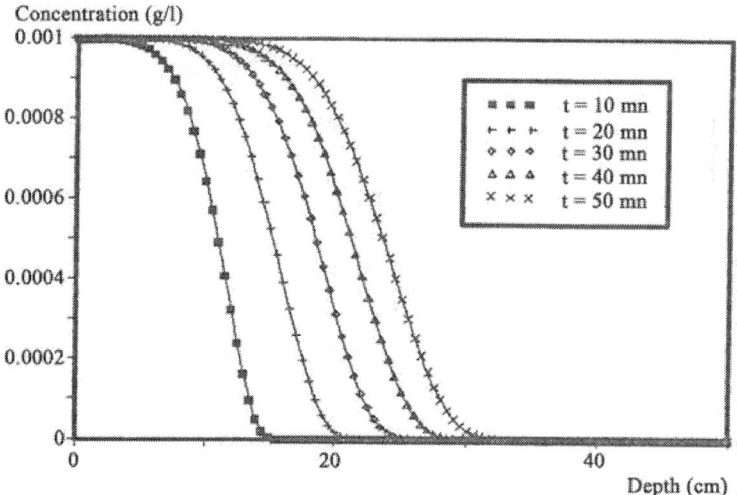

Figure 5: Pollutant concentration profiles with depth and time for a constant surface temperature.

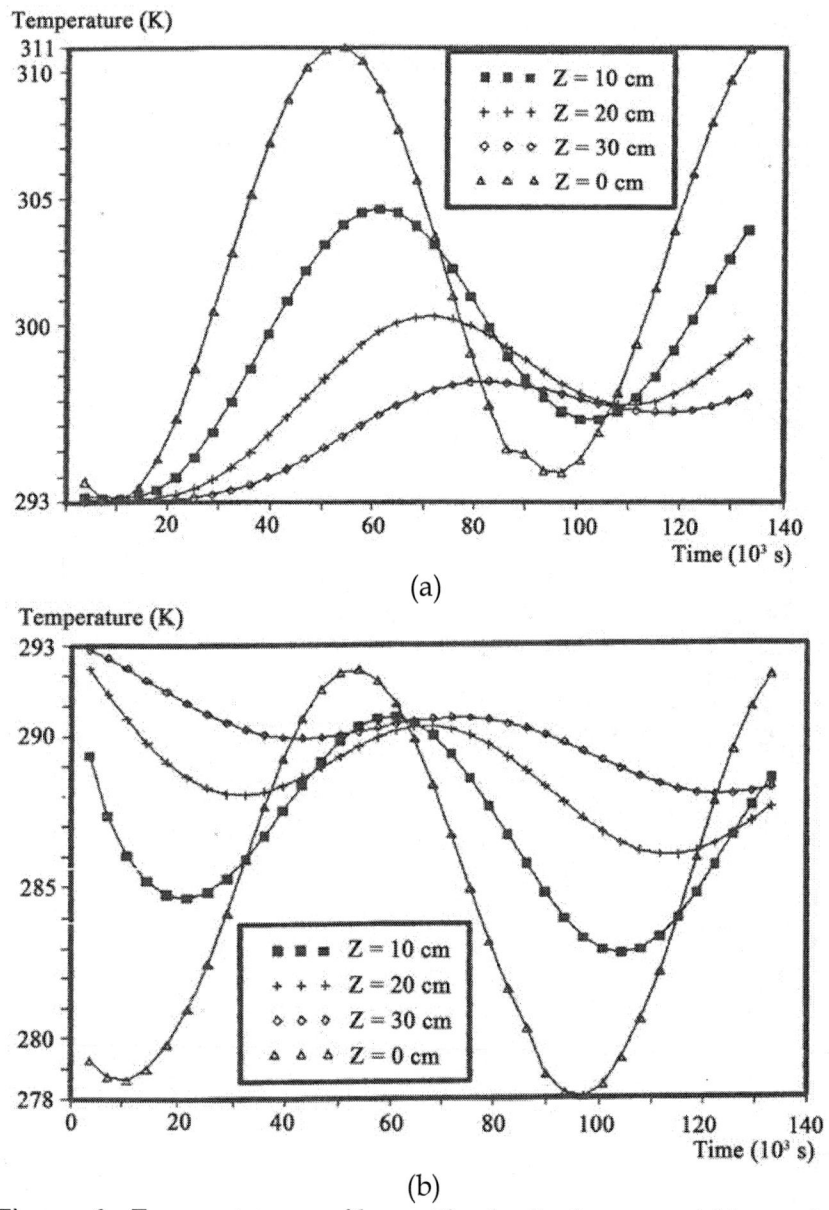

Figure 6: Temperature profiles with depth for a variable surface temperature, (a) On summer; (b) On winter.

The heat flux on the soil surface is plotted in comparison with the fluxes at various depths in Figure 7 for one day in summer and winter. Their sinusoidal form as well as the presence of a dephasing is due mainly to the condition of the variable surface temperature and the damping of the propagating thermal wave. When the ambient temperature is higher than the ground temperature, the flux is positive and the medium absorbs heat, this often occurs during the day. During the night the flux becomes negative and the ground releases heat.

For a summer day, the heat flux is mainly positive and it reaches values as high as 0.025 W/cm²during the day, and it can release 0.015 W/cm² during the night. In contrast, in winter the temperature on the surface is lower than the initial temperature of ground, its effect is thus to cool the medium. Then for a winter day the heat flux is mainly negative, with values as low as 0.04 W/cm², during the day, and it absorbs 0.01 W/cm² during the night.

The comparison of the hydrous and concentration profiles obtained in the two types of boundary conditions of temperature in one day of summer (Figure 8) shows that they are almost identical. Indeed, the diffusion of heat is very slow; its influence is thus limited to the input of the column. At the moments when computations were made, 24 h and 36 h, this part was almost saturated with water and aqueous solutions and the influence of the temperature is very small. Therefore the influence of a change in the boundary condition affects only the temperature.

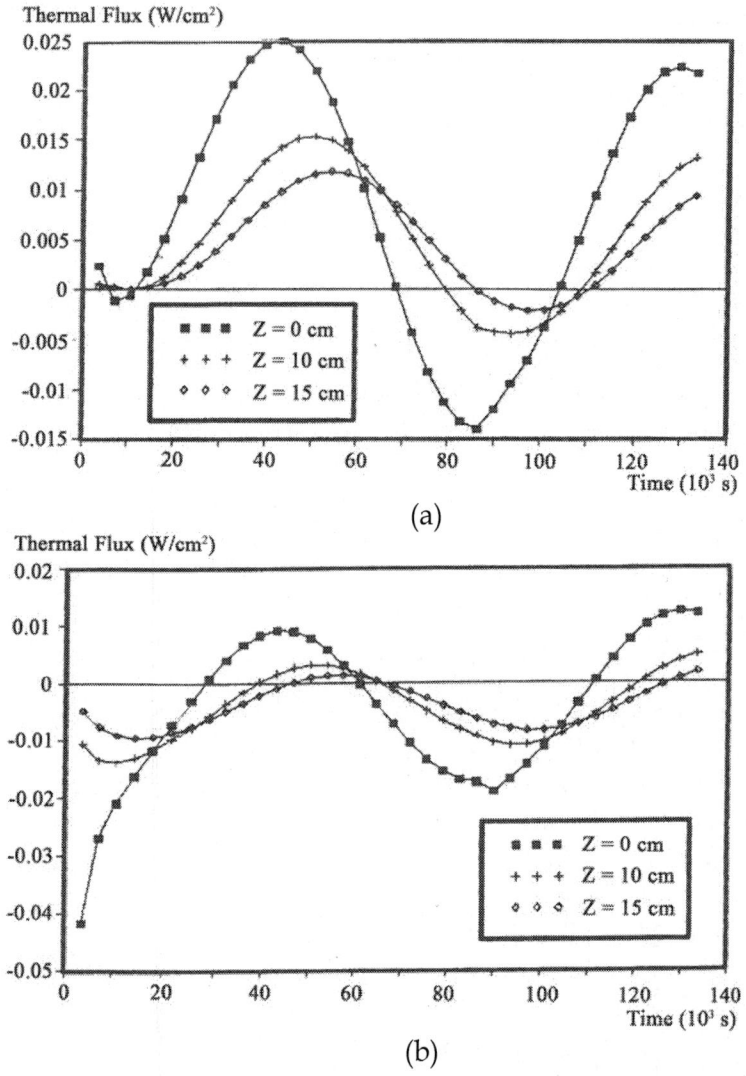

Figure 7: Heat flux profiles with depth for a variable surface temperature, (a) On summer; (b) On winter.

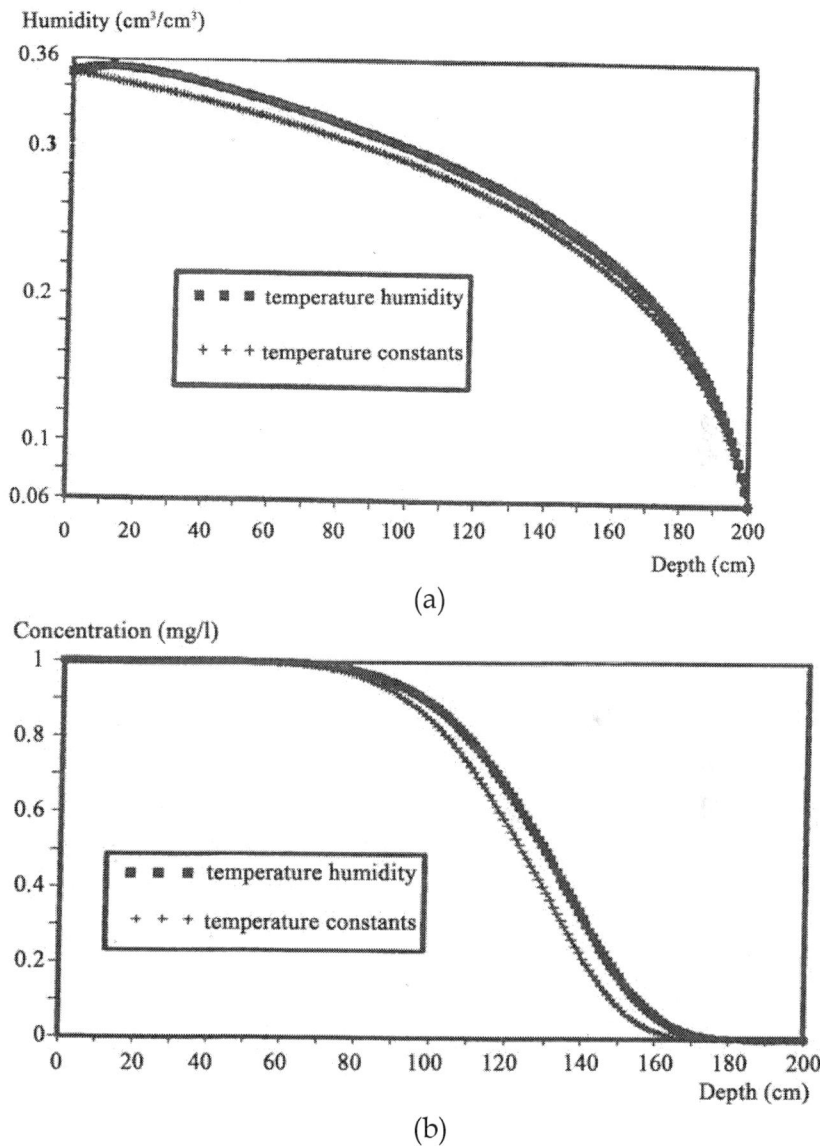

Figure 8: Comparison between curves at t = 24 h for the two kinds boundary conditions (summer). (a) Humidity; (b) Contaminant concentration.

On the other hand, for one day of winter, the temperature gradient will be directed towards outside, in opposition with the hydrous and concentration gradients, being essential on their displacement. The transfer of water and pollutant in this case will be slower than in the case of constant temperature (Figure 9).

CONCLUSIONS

The numerical study of the coupled transfers of moisture, pollutant and heat was led in the case of a monodirectional flow. This study allowed us to follow the space-time evolution of the fronts of moistening, concentration and heat, and to study the influence of change of the boundary conditions on these fronts.

The results show that the heat transfer acts on the displacement of moisture, by migration of the hot zones towards the cold zones, which in its turn modifies the profiles of concentration.

During summer, the temperature is very high and close to the constant temperature. The effect of such a change is negligible and the profiles obtained are almost the same ones. However, in winter the temperature is low and the temperature gradient is directed towards outside. The change causes a deceleration of water and pollutant transport and the variation between the curves is important.

Therefore the effect of the sinusoidal condition of the temperature on the soil surface is more important during

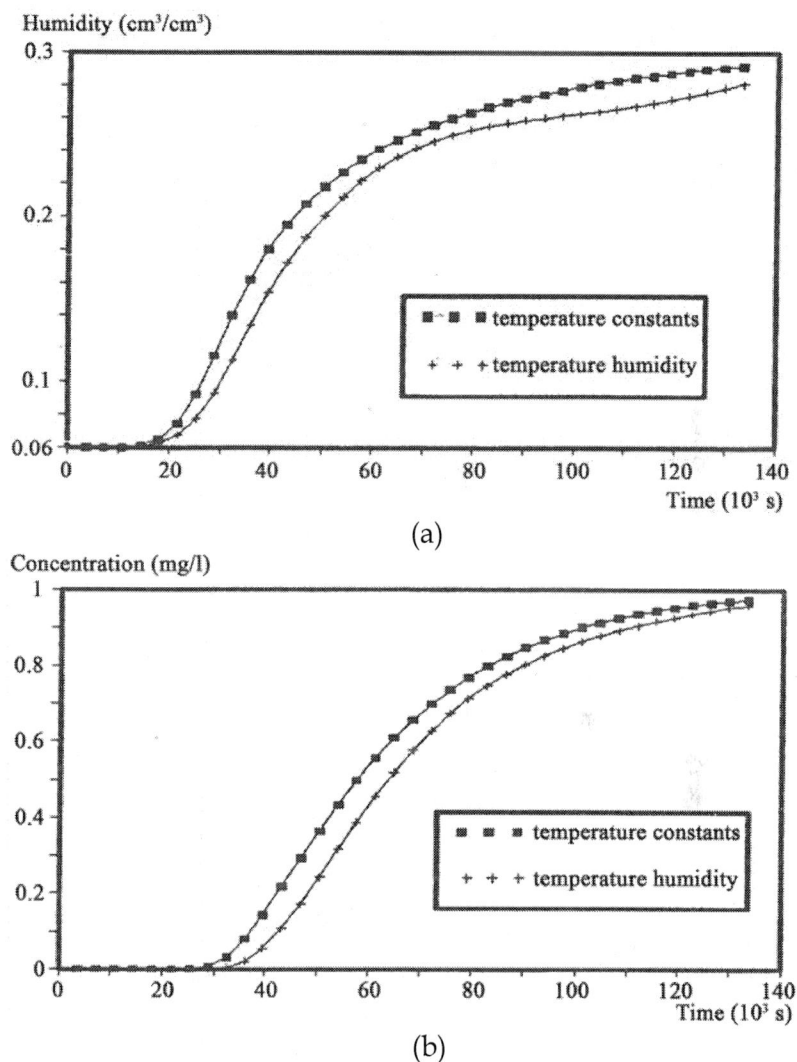

Figure 9: Comparison between temporal curves for the two kinds of boundary conditions (winter). (a) Humidity; (b) Contaminant concentration.

the winter than that during the summer, since the temperature gradient presents a great difference in both cases.

According to the results obtained, it seems that the adequate choice of the boundary conditions makes it possible to optimize the contributions of water and fertilizers in the presence of a temperature gradient.

REFERENCES

1. Gerson, H. and Mendes, N. (2006) Simultaneous Heat and Moisture Transfer in Soils Combined with Building Simulation. Energy and Buildings, 38, 303-314.
2. Jury, W.A. (1982) Simulation of Solute Transport Using a Transfer Function Model. Water Resources Research, 18, 363-368.
3. Liu, B.C., Liu, W. and Peng, S.W. (2005) Study of Heat and Moisture Transfer in Soil with a Dry Surface Layer. Intenational Journal of Heat and Mass Transfer, 48, 4579-4589.
4. Viotti, P., Petrangeli Papini, M., Stracqualursi, N. and Gamba, C. (2005) Contaminant Transport in an Unsaturated Soil: Laboratory Test and Numerical Simulation Model as Procedure for Parameters Evaluation. Ecological Modeling, 183, 131-148.http://dx.doi.org/ 10.1016/j.ecolmodel.2004.07.014
5. Jendele, L. (2002) An Improved Numerical Solution of Multiphase Flow Analysis in Soil. Advances in Engineering Software, 33, 659-668. http://dx.doi.org/10.1016/S0965-9978(02)00052-2
6. El Hadji, B.D., et al. (2001) Modeling a Non-Conservative Solute Transfer in Unsaturated Porous Media. Earth and Planetary Sciences, 333, 129-132.
7. Van Genuchten, M.T. and Wierenga, P.J. (1997) Mass Transfer Studies in Sorbing Porous Media: II. Experimental Evaluation with Tritium. Soil Science Society of America Journal, 41, 272-278. http://dx.doi. org/ 10.2136 /sssaj1977.03615995004100020022x
8. Gaudet, J.P. (1978) Transfertd'eauet de solutésdans les sols non saturés, mesures et simulation. Ph.D. Thesis, University of Grenoble, Grenoble.
9. Lewis, J. and Sjöstrom, J. (2010) Optimizing the Experimental Desing of Soil Columns in Saturated and Unsaturated Transport Experiments. Journal of Contaminant Hydrology, 115, 1-13. http://dx.doi.org/ 10.1016/j.jconhyd.2010.04.001

10. Ibnoussina, M., El Haroui, M. and Maslouhi, A. (2006) Experimentation and Modeling of the Leaching of Nitric Nitrogen in a Sanly Soil. Comptes Rendus Geoscience, 338, 787-794. http://dx.doi.org/10.1016/j.crte.2006.07.002

11. Patankar, S.V. (1978) Numerical Heat Transfer and Fluid Flow. McGraw-Hill, London.

12. Philip, J. and De Vries, D.A. (1957) Moister Movement in Porous Materials under Temperature Gradients. Transactions, American Geophysical Union, 38, 2-22.

13. Benyaich, M., Belghit, A., Gonzalez, A. and Claudet, B. (1998) Influence of Temperature on the Transport of Pollutant in Non-Saturated Soil. 3rd International Conference on Hydroscience and Engineering, Brandenburg University of Technology at Cottbus, ICHE, Cottbus/Berlin, 31 August-3 September 1998.

14. Crausse, P. (1982) Etude fondamentale des transferts de chaleuretd'humidité. en milieu poreux non saturé. P.h.D., I.N.P. Toulouse.

15. Ouardi, C. and Zrikem, Z. (1997) Sélection de séquencestypiques pour le climat de Marrakech: Application à l'habitat. 3ème Congrès de Mécanique, Tome II-b, Tetouan, 673-679.

16. Pickens, J.F., Gilham, R.W. and Cameron, D.R. (1979) Finite Element Analysis of the Transport of Water and Solutes in the Drained Soils. Journal of Hydrology, 40, 243-264.http://dx.doi.org/10.1016/0022-1694(79)90033-7

Nomenclature

θ: Volumetric Water Contents

q_θ: Flux Density of Water

$K(\theta)$: Hydraulic conductivity

D_θ: Isothermal Water Diffusivity

D_T: Non-isothermal Water Diffusivity

Dap: Coefficient of Hydrodynamic Dispersion

C: Concentration of pollutant

λ: Thermal Conductivity

C_p: Volumetric Heat Capacity of the Soil

T: Temperature

z: Space Coordinate

t: Time Coordinate

CITATION

Belghit, A. and Benyaich, M. (2014) Numerical Study of Heat Transfer and Contaminant Transport in an Unsaturated Porous Soil. Journal of Water Resource and Protection, 6, 1238-1247. doi: 10.4236/jwarp.2014.613113.

CHAPTER 4

Experimental Determination of Heat Transfer Coefficients During Squeeze Casting of Aluminium

Jacob O. Aweda [1] and Michael B. Adeyemi[1]

[1] Department of Mechanical Engineering, University of Ilorin, Ilorin, Nigeria

INTRODUCTION

Casting process is desired because it is very versatile, flexible, and economical and happens to be the shortest and fastest way to transform raw material into finished product. Squeeze casting belongs to permanent mould casting method which offers considerable saving in cost for large production quantities when the size of the casting is not large. Squeeze casting has the advantage of producing good surface finish, close dimensional tolerance and the absence of sand inclusions on the cast surfaces of the products as opined by Das and Chatterjee, (1981).

The solidification process of the molten aluminium metal in the steel mould takes a complex form, (Hosford and Caddell, 1993) and (Potter and Easterling, 1993). During solidification all mechanisms of heat transfer are involved and the solidifying metal undergoes state and phase changes. The final structure and properties of the cast product obtained depend on the casting parameters applied i.e. applied pressures, die pre-heat temperature, delay time and period of applied pressure on the solidifying metal, (Potter and Easterling, 1993), (Bolton, 1989) and (Callister, 1997). The prediction of temperature distribution and solidification rate in metal casting is very important in modern foundry technologies. This helps to control the fundamental parameters such as the occurrence of defects, as well as, the influence on final properties of cast products and the mould wall / cast metal interface contact surface.

Heat transfer coefficients during squeeze cast of commercial aluminium were determined using the solidification temperature versus time curves obtained for varying applied pressures during squeeze casting process. The steel mould / cast aluminium metal interface temperatures versus times curve obtained through polynomial curves fitting and extrapolation was compared with the numerically obtained temperatures versus times curve. Interfacial heat transfer coefficients were determined experimentally from measured values of heating and cooling temperatures of steel mould and cast metal and compared with the numerically obtained values and found to be fairly close in values.

Aluminium is a product with unique properties, making it a natural partner for the building and other manufacturing industries.The commercially pure aluminium metal used for this research work finds extensive use in the building, manufacturing and process industries, both as a material of construction and household goods. Products of squeeze casting are of improved mechanical properties and could be given heat treatment. Heat dissipation from the squeeze cast specimen is fast thus producing products of fine grains as compared to the slow cooling of sand casting, which produces large grains. Products obtained through squeeze casting are with improved mechanical properties.

SQUEEZE CASTING PROCEDURE

A metered quantity of molten metal was poured into the steel mould cavity at a supper-heat temperature of between 40-60 ^{0}C fast but avoiding turbulence. The upper die was then released to close the mould cavity with and without applying any load on the upper die. Thermocouples were inserted into the drilled holes made in the die, which were used to monitor both the die and cast metal temperatures. The terminals of the thermocouples were connected to the chart recorder/plotter (set at the highest speed of 10mm/s and voltage 100mV) through the cold junction apparatus, maintained at 0 ^{0}C throughout the measuring period.

ASSUMPTIONS MADE

i. Heat transfer in the molten metal cast zone is due to both conduction and convection while conduction heat transfer occurs in the steel mould, it is convection at the outer surface of the steel mould.

ii. The thickness of the cast specimen is much smaller than the diameter (radial dimension), thus giving one dimensional heat transfer process.

iii. Considering the symmetrical nature of the cast specimen, solidification process was assumed symmetrical and only lower half of the specimen's thickness was analysed see figure 1.

iv. The bottom of the squeeze casting rig was and the heat losses to the atmosphere was small and neglected.

v. Heat losses through conduction and convection to the atmosphere at the punch were neglected, as a result of short time of pressure application.

vi. The process of analyses in the cast specimen starts only when the steel mould cavity had been filled with the required quantity of liquid molten metal (i.e. heat transfer processes during pouring of molten aluminium into the steel mould are not considered).

vii. Density of the molten and solidified aluminium metal was assumed to be the same and independent of temperature.

viii. Thermal conductivity and specific heat of aluminium metal were dependent on the cast temperatures.

HEAT TRANSFER GOVERNING EQUATIONS

Without Pressure Application on the Cast Metal

A measured quantity of molten aluminium metal was poured into the steel mould cavity. The process of solidification begins from the steel mould/cast metal interface and continues inwards into the cast metal. As this process continues, there was an increase in the thickness of the solidified layer and a decrease in the liquid molten metal portion. For the situation when nopressure was applied on the solidified molten metal, the governing heat transfer equations in one dimension are given by equation (1).

$$\rho C \frac{\partial T}{\partial t} = K \left[\frac{\partial^2 T}{\partial r^2} + \frac{1}{r}\frac{\partial T}{\partial r} \right] \tag{1}$$

From figure 1, equation (1) is defined within the region with;

i. Steel mould,

$$L \leq r_{st} \leq (L + Q)$$

ii. Solidified molten,

$$\left(L - X^j_r\right) \leq r_s \leq L$$

$$T_S = T_{MM} = 660 \left(^0 C\right) \tag{2}$$

iii. Liquid molten metal,

$$0 \leq r_L \leq \left(L - X^j_r\right)$$

$$K_S \frac{\partial T_S}{\partial r} = 0 \tag{3}$$

$$r = 0 \tag{4}$$

$$T_L = T_P = 720 \left(^0 C\right) \tag{5}$$

iv. At the phase change boundary condition;

$$\rho_L L_f \frac{dX^j_r}{dt} = K_L \frac{\partial T_L}{\partial r} - K_S \frac{\partial T_S}{\partial r} \tag{6}$$

where,

$$r = L - X^j_r \tag{7}$$

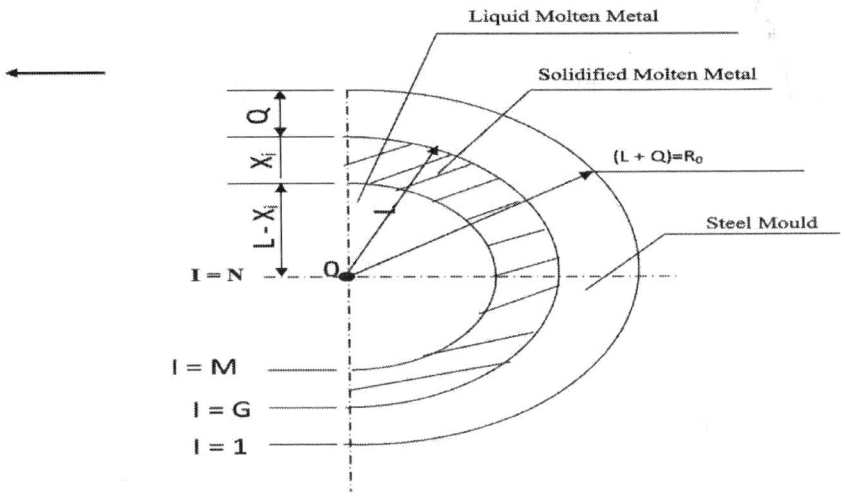

Figure 1: Schematic representation of solidification front in one dimension (radial direction).

Casting with Pressure Application on the Solidified Molten Metal

As the cast aluminium metal solidifies, pressure is applied on the specimen, observing lapse or delay time, t while varying the values of pressure applied. The time between the end of pouring of molten metal and pressure application known as lapse time, is recorded. This is necessary such that the cast specimen will not stick to the upper punch or cause the cast metal to tear with pressure application. Due to the applied pressure, an internal energy Δq is generated within the solidified molten metal, (see figure 2).

Inserting the internal energy into the heat transfer equation (1) for solidified molten metal, it becomes equation (8),

$$\rho_s C_s \frac{\partial T_s}{\partial t} = K_s \left[\frac{\partial^2 T_s}{\partial r^2} + \frac{1}{r_s} \frac{\partial T_s}{\partial r} \right] + \Delta q \tag{8}$$

where,

$$\nabla q = \nabla q_p + \nabla q_f \tag{9}$$

Δq-internal energy generated by applied pressure,

Δq$_P$-energy due to plastic strain within the solidified molten metal material,

Δq$_f$-frictional energy generated during pressure application,

$$\nabla q_f = \nabla q_{fP} + \nabla q_{fm} \tag{10}$$

Δq$_{fP}$-frictional energy due to punch / solidified molten metal interface,

Δq$_{fm}$-frictional energy due to steel mould cylindrical surface / solidified molten metal interface.

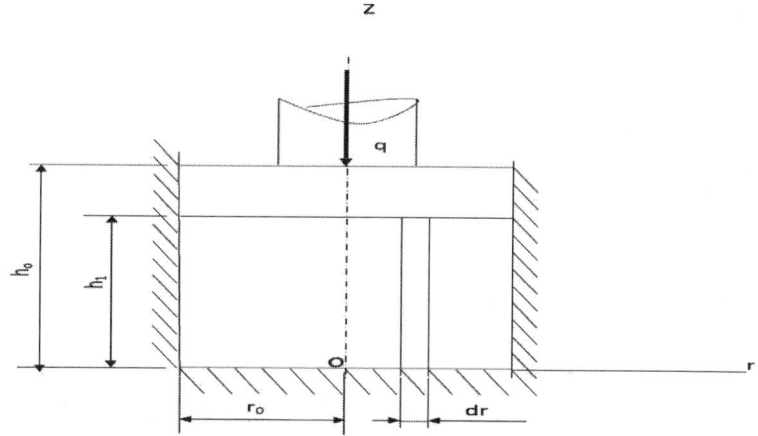

Figure 2: Cast specimen under pressure.

FORMULATION OF HEAT TRANSFER EQUATIONS

Finite difference expressions for the nodal temperatures are obtained by either energy balance within an elemental volume around the node or by substitutions into the governing partial differential equations. Partial time derivative of temperature contained in the equations can be written in terms of a moving gradient as inWhite (1991) at a velocity of dx/dt. In the heat transfer equation, temperature is defined as a function of distance, r with time, t and represented in equation (11).

$$T = T(r, t) \tag{11}$$

In the Steel Mould

$$\frac{dT_{st}}{dt} = \alpha_{st} \frac{\partial^2 T_{st}}{\partial r^2} + \frac{\alpha_{st}}{r_{st}} \frac{\partial T_{st}}{\partial r} \tag{12}$$

where,

$$\frac{dr_{st}}{dt} = 0 \tag{13}$$

$$\alpha_{st} = \frac{K_{st}}{\rho_{st} C_{st}} \tag{14}$$

st-thermal diffusivity of steel mould material.

In the Solidified Molten Metal

$$\frac{dT_S}{dt} - \frac{(1-G)}{(M-G)} \frac{dX_i^r}{dt} \frac{\partial T_S}{\partial r} = \alpha_S \frac{\partial^2 T_S}{\partial r^2} + \frac{\alpha_S}{r_S} \frac{\partial T_S}{\partial r} \tag{15}$$

where,

$$\alpha_S = \frac{K_S}{\rho_S C_S} \tag{16}$$

s-thermal diffusivity of solidified molten metal material.

In the Liquid Molten Metal

$$\frac{dT_L}{dt} - \frac{(I-M)}{(N-M)} \frac{dX_i^r}{dt} \frac{\partial T_L}{\partial r} = \alpha_L \frac{\partial^2 T_L}{\partial r^2} + \frac{\alpha_L}{r_L} \frac{\partial T_L}{\partial r} \tag{17}$$

where,

$$\alpha_L = \frac{K_L}{\rho_L C_L} \tag{18}$$

$_L$ -thermal diffusivity of liquid molten metal material.

NODAL DIVISIONS

The steel mould, the solidified metal and the molten metal regions were discretized separately. Each of these regions was divided into a fixed number of gridal points as in figure 1.

In the Steel Mould

$$d_{st} = \frac{Q}{(G-I)} \tag{19}$$

I = 1, 2, 3,..., G-1

In the Solidified Molten Metal

$$d_S = \frac{X_r^i}{(M-I)} \tag{20}$$

I = G+1, G+2, G+3, ..., M-1

In the Liquid Molten Metal Portion

$$d_L = \frac{\left(L - X_r^i\right)}{(N-I)} \tag{21}$$

$$I = M+1, M+2, M+3 \ldots N-1$$

In the Phase Change Boundaries

The phase change is represented by the equations;

$$d_{Ps} = \frac{X_r^i}{(M-G)} \qquad (22)$$

$$d_{PL} = \frac{(L - X_r^i)}{(N-M)} \qquad (23)$$

At the Completion of Solidification

$$d_{SC} = \frac{L}{(N-G)} \qquad (24)$$

As solidification time progresses, the boundary locations change, and the thickness of the solidified molten metal in the radial direction increases. The rate of change of boundary location with time is represented, mathematically in equation (25),

$$\frac{dX_r^i}{dt} = \frac{(X_r^{j+1} - X_r^j)}{\delta} \qquad (25)$$

Where,

$$(X_r^{j+1} - X_r^j)$$

-are differences in the thickness of solidified molten metal at a particular time interval as time progresses in the radial direction.

δ -time interval.

BOUNDARY CONDITIONS

The problem of phase change during solidification is that the location of the solidifying molten metal / liquid molten metal interface is not known and this is determined continuously by appropriate mathematical analysis. This moving interface is normally expressed mathematically by the energy balance equations at the interfaces. In the numerical analysis, as solidification of molten aluminium metal progresses, three boundary interfaces occurred as:

Steel Mould-Atmosphere Interface (I = 1)
The heat conducted to the steel mould material (from I = 2 to 1 = 1) equals the sum of the change in the internal energy and heat convected from the surface of the steel metal mould material into the atmosphere, mathematically represented in equation (26).

$$\frac{\partial T_{st}}{\partial t} = \frac{2K_{st}}{\rho_{st} C_{st} d_{st}} \frac{\partial T_{st}}{\partial r} - \frac{2H^*}{\rho_{st} C_{st} d_{st}} (T_i^j - T_\infty) \tag{26}$$

$$I = 1$$

In the Solidified Molten Metal-Steel Mould Interface (I= G)
The sum of the heat conducted from the solidifying molten metal/steel mould interface and the change (decrease) in the internal energy at the boundary equal the sum of the heat conducted to the steel mould and the change (increase) in the internal energy at the interface.

$$K_S \frac{\partial T_S}{\partial r} + \frac{1}{2}\rho_S C_S d_S \frac{\partial T_S}{\partial t} = K_{st} \frac{\partial T_{st}}{\partial r} + \frac{1}{2}\rho_{st} C_{st} d_{st} \frac{\partial T_{st}}{\partial t} \tag{27}$$

$$I = G$$

In the Liquid Molten Metal – Solidified Molten Metal Interface, (I= M)

The sum of the heat conducted from the liquid molten metal and the internal energy generated equal the sum of heat conducted to the solidified molten metal and the internal energy generated at the interface as inequation (28).

$$K_L \frac{\partial T_L}{\partial r} - \frac{1}{2}\rho_L C_L d_L \frac{\partial T_L}{\partial t} = K_S \frac{\partial T_S}{\partial r} + \frac{1}{2}\rho_S C_S d_S \frac{\partial T_S}{\partial t} \qquad (28)$$

I = M

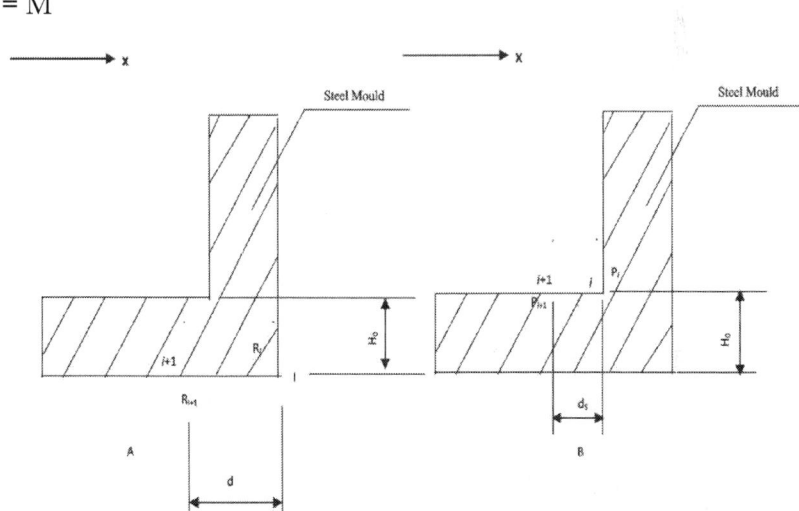

Figure 3: Corner nodes. A) External, B) Internal nodes

At the Corners

The usual one dimensional heat transfer analysis does not take into consideration heat loss at the corners as represented in figure 3. Heat loss at both the external and internal corner nodes of the steel mould have been considered and analysed while the die is lagged at the bottom surface.

External Corner Effect

$$q_{i+1} + q_g = q_{conv} \tag{29}$$

The heat conducted from point R_{i+1} to point R_i added to the change in the internal energy equal the amount of heat convected out to the atmosphere at point R_I., Fig. (3a). The finite form of equation (29) is represented by equation (30),

$$\frac{K_{st}}{d_{st}}\frac{\partial T_{st}}{\partial r} - \frac{1}{2}\rho_{st}C_{st}\frac{\partial T_{st}}{\partial t} = \frac{h}{d_{st}}\left(T_i^j - T_\infty^j\right) + \frac{h}{H}\left(T_i^j - T_\infty^j\right) \tag{30}$$

Internal Corner Effect

The heat conducted from point P_{I+1} to point P_I added to the change in the internal energy amount to the heat conducted into the steel mould at point I, fig. (3b) thus becoming equation (31);

$$\frac{K_S}{d_{st}}\frac{\partial T_S}{\partial r} + \frac{1}{2}\rho_S C_S\frac{\partial T_{st}}{\partial t} = \frac{K_{st}}{d_S}\frac{\partial T_{st}}{\partial r} + \frac{K_{st}}{H1}\frac{\partial T_{st}}{\partial Z} \tag{31}$$

First Time Analysis

Solidification takes place only in the radial direction, a one-dimensional heat solidification problem was assumed numerically to take place in the radial direction only after filling the steel mould cavity with the liquid molten metal.

For the first time analysis, the specimen is considered to be in the molten stage and therefore equation (32) for the liquid molten metal is used for computation. Thus;

$$\rho_L C_L \frac{dT_L}{dt} = K_L\left[\frac{\partial^2 T_L}{\partial r^2} + \frac{1}{r_L}\frac{\partial T_L}{\partial r}\right] \tag{32}$$

This equation is subjected to the boundary conditions with;

$$T_L = T_S \tag{33}$$

The instantaneous radius ri in the first time analysis is given by equation (34);

$$r_L = \frac{(N-I)}{(N-G)} L \tag{34}$$

I = G, G+1, G+2, G+3,...., N

The boundary velocity of moving coordinate of equation (32) is given as equation (35);

$$\frac{dr_i}{dt} = \frac{d}{dt}\left[\frac{(N-I)}{(N-G)} L\right] = 0 \tag{35}$$

At the Completion of Solidification

At the completion of solidification, the whole molten region becomes solidified molten metal, and just before pressure is applied, the governing heat transfer equation becomes equation (36) representing the solidified portion;

$$\rho_S C_S \frac{\partial T_L}{\partial t} = K_S\left[\frac{\partial^2 T_S}{\partial r^2} + \frac{1}{r_{SC}} \frac{\partial T_S}{\partial r}\right] \tag{36}$$

The equation is applicable within the region defined by equation (37);

$$r = L \tag{37}$$

The boundary motion at the completion of solidification is as in equation (38);

$$r_{SC} = \frac{(N-I)}{(N-G)} L \tag{38}$$

I = G, G+1, G+2, G+3,...,.N

The boundary velocity at the completion of solidification is expressed as equation (39);

$$\frac{dr_{SC}}{dt} = \frac{d}{dt}\left[\frac{(N-I)}{(N-G)}L\right] = 0 \tag{39}$$

where,

L -constant value

FINITE DIFFERENCE OF GOVERNING HEAT TRANSFER EQUATIONS

The heat transfer equations generated in the cast metal and interfaces are written in the finite difference forms. These equations are presented in the various regions thus;

In the Steel Mould Region

$$T_i^{j+1} = \left[1 - \frac{2\delta\alpha_{st}}{d_{st}^2} - \frac{\delta\alpha_{st}}{r_{st}d_{st}}\right]T_i^j + \left[\frac{\delta\alpha_{st}}{d_{st}^2} + \frac{\delta\alpha_{st}}{r_{st}d_{st}}\right]T_{i+1}^j + \frac{\delta\alpha_{st}}{d_{st}^2}T_{i-1}^j \tag{40}$$

where,

$$r_{st} = L + \frac{Q(G-I)}{(G-1)} \quad \text{(a)}$$

$$d_{st} = \frac{Q}{(G-I)} \quad \text{(b)} \tag{41}$$

I = 1,2,3,..............G

In the Solidified Molten Metal Portion

$$T_i^{j+1} = \left[1 - \frac{(I-G)}{(M-G)}\frac{\left(X_r^{j+1} - X_r^j\right)}{d_S} - \frac{2\delta\alpha_S}{d_S^2} - \frac{\delta\alpha_S}{r_S d_S}\right]T_i^j +$$

$$+ \left[\frac{(I-G)}{(M-G)}\frac{\left(X_r^{j+1} - X_r^j\right)}{d_S} + \frac{\delta\alpha_S}{d_S^2} + \frac{\delta\alpha_S}{r_S d_S}\right]T_{i+1}^j + \frac{\delta\alpha_S}{d_S^2}T_{i-1}^j$$

(42)

where,

$$r_S = L - \frac{X_r^j(I-G)}{(M-G)} \quad \text{(a)}$$

$$d_S = \frac{X_r^j}{(M-I)} \quad \text{(b)}$$

(43)

$$I = G+1,\ G+2,\ G+3, \ldots\ldots\ldots, M$$

In the Liquid Molten Metal Region

$$T_i^{j+1} = \left[1 - \frac{(I-M)}{(N-M)}\frac{\left(X_r^{j+1} - X_r^j\right)}{d_L} - \frac{2\delta\alpha_L}{d_L^2} - \frac{\delta\alpha_L}{r_L d_L}\right]T_i^j +$$

$$\left[\frac{(I-M)}{(N-M)}\frac{\left(X_r^{j+1} - X_r^j\right)}{d_L} + \frac{\delta\alpha_L}{d_L^2} + \frac{\delta\alpha_L}{r_L d_L}\right]T_{i+1}^j + \frac{\delta\alpha_L}{d_L^2}T_{i-1}^j$$

(44)

where,

$$r_L = \left(L - X_r^j\right)\frac{(I-M)}{(N-M)} \quad \text{(a)}$$

$$d_L = \frac{\left(L - X_r^j\right)}{(N-I)} \quad \text{(b)}$$

(45)

$$I = M, M+1, M+2, M+3, \ldots, N$$

In the Phase Change Boundary Condition (I = M)

$$X_r^{j+1} = X_r^j - \left[\frac{\delta K_L}{\rho_L L_f h_{pL}} + \frac{\delta K_S}{\rho_L L_f h_{ps}} \right] T_i^j + \frac{\delta K_L}{\rho_L L_f h_{pL}} T_{i+1}^j + \frac{\delta K_S}{\rho_L L_f h_{ps}} T_{i-1}^j \qquad (46)$$

where,

$$h_{PL} = \frac{\left(L - X_r^j \right)}{\left(N - M \right)} \qquad (a)$$

$$h_{Ps} = \frac{X_r^j}{\left(M - G \right)} \qquad (b)$$

$$(47)$$

In the Steel Mould / Atmosphere Interface (I = 1)

$$T_i^{j+1} = \left[1 - \frac{2\delta K_{st}}{d_{st}^2 \rho_{st} C_{st}} \right] T_i^j + \frac{2\delta K_{st}}{d_{st}^2 \rho_{st} C_{st}} T_{i+1}^j - \frac{2\delta H^*}{d_{st} \rho_{st} C_{st}} \left(T_i^j - T_\infty^i \right) \qquad (48)$$

I = 1

where;

$$d_{st} = \frac{Q}{\left(G - 1 \right)} \qquad (49)$$

I = 1, 2, 3, …, G

In the Solidified Molten Metal / Steel Mould Interface (I = G)

$$T_i^{j+1} = a\left[\frac{2\delta K_S}{d_S} + d_S\rho_S C_S + \rho_S C_S\frac{(I-G)}{(M-G)}\left(X_r^{j+1} - X_r^j\right) + \frac{2\delta K_{st}}{d_{st}} - d_{st}\rho_{st} C_{st}\right]T_i^j$$

$$+a\left[\rho_S C_S\frac{(I-G)}{(M-G)}\left(X_r^{j+1} - X_r^j\right) - \frac{2\delta K_S}{d_S}\right]T_{i+1}^j - \frac{2a\delta K_{st}}{d_{st}}T_{i-1}^j$$

(50)

where,

$$a = \frac{1}{\left(d_S\rho_S C_S - d_{st}\rho_{st} C_{st}\right)} \quad \text{(a)}$$

$$d_{st} = \frac{Q}{(G-I)} \quad \text{(b)}$$

$$I = 1, 2, 3,..., G$$

(51)

$$d_S = \frac{X_r^i}{(M-I)} \quad \text{(c)}$$

$$I = G+1, G+2, G+3,..., M$$

In the Liquid Molten Metal / Solidified Molten Metal Interface (I = M)

$$T_i^{j+1} = b\left[\begin{matrix}d_L\rho_L C_L - \rho_L C_L\frac{(I-M)}{(N-M)}\left(X_r^{j+1} - X_r^j\right) - \frac{2\delta K_L}{d_L} + d_S\rho_S C_S - \frac{2\delta K_S}{d_S} + \\ \rho_S C_S\frac{(I-G)}{(M-G)}\left(X_r^{j+1} - X_r^j\right)\end{matrix}\right]T_i^j$$

(52)

$$+b\left[\frac{2\delta K_S}{d_S} - \rho_S C_S\frac{(I-G)}{(M-G)}\left(X_r^{j+1} - X_r^j\right)\right]T_{i-1}^j$$

$$+b\left[\frac{2\delta K_L}{d_L} + \rho_L C_L\frac{(I-M)}{(N-M)}\left(X_r^{j+1} - X_r^j\right)\right]T_{i+1}^j$$

where,

$$b = \frac{1}{\left(d_S \rho_S C_S + d_L \rho_L C_L\right)} \quad \text{(a)}$$

$$d_S = \frac{X_r^j}{(M-1)} \quad \text{(b)}$$

$$d_L = \frac{\left(L - X_r^j\right)}{(N-1)} \quad \text{(c)}$$

(53)

$$I = M, M+1, M+2, M+3,\ldots,N$$

External Corner Effect (I = 1)

$$T_i^{j+1} = \left[1 - \frac{2\delta K_{st}}{\rho_{st}C_{st}d_{st}^2} - \frac{2\delta h}{\rho_{st}C_{st}d_{st}} - \frac{2\delta h}{\rho_{st}C_{st}H}\right]T_i^j +$$

$$+ \frac{2\delta K_{st}}{\rho_{st}C_{st}d_{st}^2}T_{i+1}^j +$$

$$+ \left[\frac{2\delta h}{\rho_{st}C_{st}d_{st}} + \frac{2\delta h}{\rho_{st}C_{st}H}\right]T_\infty^j$$

(54)

where,

$$d_{st} = \frac{Q}{(G-1)}$$

(55)

Internal Corner Effect (I = G)

$$T_i^{j+1} = \left[1 + \frac{2\delta K_S}{\rho_S C_S d_S^2} + \frac{(I-G)}{(M-G)}\frac{\left(X_r^{j+1} - X_r^j\right)}{d_S} + \frac{2\delta K_{st}}{\rho_S C_S d_S d_{st}} + \frac{2\delta K_{st}}{\rho_S C_S H1 d_{st}}\right]T_i^j -$$

(56)

$$- \frac{2\delta K_S}{\rho_S C_S d_S^2}T_{i+1}^j - \left[\frac{2\delta K_{st}}{\rho_S C_S d_S d_{st}} + \frac{2\delta K_{st}}{\rho_S C_S H1 d_{st}}\right]T_{i-1}^j$$

where,

$$d_S = \frac{X_r^j}{(M-I)} \qquad \text{(a)}$$

$$I = G+1, G+2, G+3, \ldots, M \qquad (57)$$

$$d_{st} = \frac{Q}{(G-I)} \qquad \text{(b)}$$

$$I = 1, 2, 3, \ldots, G$$

First Time Analysis

Finite difference form of equation (32) therefore becomes equation (58);

$$T_i^{j+1} = \left[1 - \frac{2\delta\alpha_L}{d_L^2} - \frac{\delta\alpha_L}{r_L d_L}\right]T_i^j + \left[\frac{\delta\alpha_L}{d_L^2} + \frac{\delta\alpha_L}{r_L d_L}\right]T_{i+1}^j + \frac{\delta\alpha_L}{d_L^2}T_{i-1}^j \qquad (58)$$

where,

$$r_L = \frac{(N-I)}{(N-G)}L \qquad \text{(a)}$$

$$d_L = \frac{L}{(N-G)} \qquad \text{(b)}$$

$$(59)$$

$$I = G+1, G+2, G+3, \ldots\ldots\ldots, N$$

At the Completion of Solidification

Finite difference form of equation (36) becomes equation (60);

$$T_i^{j+1} = \left[1 - \frac{2\delta\alpha_S}{d_{SC}^2} - \frac{\delta\alpha_S}{r_{SC}d_{SC}}\right]T_i^j + \left[\frac{\delta\alpha_S}{d_{SC}^2} + \frac{\delta\alpha_S}{r_{SC}d_{SC}}\right]T_{i+1}^j + \frac{\delta\alpha_S}{d_{SC}^2}T_{i-1}^j \qquad (60)$$

where;

$$r_{SC} = \frac{(N-I)}{(N-G)} L \quad \text{(a)}$$

$$d_{SC} = \frac{L}{(N-G)} \quad \text{(b)}$$

(61)

$$I = G+1, G+2, G+3,..., (N-1)$$

STABILITY CRITERIA

For stability criteria to be achieved, the values of temperature

$Tj'i$

in all the heat governing equations should not be negative according to Ozisik (1985) and White (1991) not to negate the law of thermodynamics which could lead to temperature fluctuations.Therefore, for stability to be achieved, the coefficients of

$Tj'I$ in each of the equations must be greater than zero.

CASTING WITH PRESSURE APPLICATION AND DIE HEATING

Pressure was applied only when the cast specimen was solidified, the governing heat transfer equation therefore, takes the form of solidified molten metal (completion of solidification). The finite difference ofequation (8) is written as equation (62);

$$T_i^{j+1} = \left[1 - \frac{2\delta\alpha_S}{d_S^2} - \frac{\delta\alpha_S}{r_S d_S} \right] T_i^j + \left[\frac{\delta\alpha_S}{d_S^2} + \frac{\delta\alpha_S}{r_S d_S} \right] T_i^j + \frac{\delta\alpha_S}{d_S^2} T_{i-1}^j + \Delta T$$

(62)

where,

T -temperature change resulting from pressure application

The cast specimen height, h_c, is pressure dependent and the relationship is expressed with the equation (63) below after performing series of experiment with various applied pressure;

$$h_c = -0.00007P + 0.036833 \qquad (63)$$

where coefficient of correlation r = 0.996

The plastic flow stress, $\sigma_{(T)}$, is dependent on both the applied pressure, P, and die temperature, TM,(White, 1991), and expressed with the equation (64);

$$\sigma(T) = 0.244P - 0.0405TM + 18.614 \qquad (64)$$

where coefficient of correlation r = 0.9508

CASTING WITH DIE HEATING

Aluminium cast specimens were produced with die pre- heating temperatures of between 100- 300°C without applying pressure on the solidifying aluminium metal. The die heating process was carried out, using three electric heater rods (100Watts each) that were connected to a.c supply. The required die temperatures were set and controlled, using a bimetallic thermostat.

HEAT TRANSFER COEFFICIENT EVALUATIONS

The method of calculating heat transfer coefficients as reported by Santos et al (2001) and Maleki et al, (2006) is based on the knowledge of known temperature histories at the interior points of the casting or mould together with the numerical models of heat flow during solidification. These temperatures are difficult to measure due to the difficulty in locating accurate position of thermocouple at the interface. Therefore, the inverse heat conduction problem based on non-linear estimation technique of Chattopadhyay, (2007) and Hu and Yu,(2002), has been adopted to determine the values of interface heat transfer coefficients, as a function of time during solidification of squeeze casting. Solidification of squeeze casting of aluminium involves phase change and therefore thermal properties of aluminium are temperature dependent, making the inverse heat conduction problem non-linear.

The governing heat transfer equation in one-dimensional cylindrical coordinates is given by equation (65):

$$\rho c_p \frac{\partial T}{\partial t} = \frac{1}{r} \frac{\partial}{\partial r}\left(K_{al} r \frac{\partial T}{\partial r}\right) \tag{65}$$

Equation (65) holds within the boundary condition as expressed in equation (66):

$$q = K_{al}(T)\frac{\partial T}{\partial r} = h_{al}(T)\left[T_{al} - T_M \right] \tag{66}$$

The thermal conductivity K_{al} (Reed-Hill and Abbaschian 1973) and (Elliot, 1988) of aluminium is dependent upon casting temperature, T_{al}, and expressed in equation (67):

$$K_{al}(T) = 241.84 - 0.041 T_{al} \tag{67}$$

The heat flow across the casting/mould interface can be characterized by an average interfacial heat transfer coefficient, $h_{al}(T)$ as obtained by Gafur et al (2003) and Santos et al (2004). This is expressed mathematically in equation (68):

$$h_{al}(T) = \frac{q}{\left[T_{al} - T_M\right]} \qquad (68)$$

The heat transfer coefficient, h, at the interface is estimated by minimizing the errors between numerically estimated and measured temperatures defined by equation (69):

$$F(h) = \sum_{i=1}^{n}\left(T_{est} - T_{exp}\right)^2 \qquad (69)$$

where,

T_{est} and T_{exp} -are the estimated and experimentally measured temperatures at various thermocouples location and times,

n -iteration stage

NUMERICAL SIMULATIONS OF DIFFERENTIAL EQUATIONS

Squeeze casting consists of two stages, the first of which is mould filling: - the mould is filled with the required quantity of liquid molten metal; the second is cooling, this continues until the part has solidified completely. Controlling both stages is of major importance for obtaining sound casts with the required geometry and mechanical properties as observed by (Kobryn and Semiatin, (2000), Browne and O'Mahoney,(2001) andMartorano and Capocchi,(2000). When molten metal is poured into the mould cavity, it is initially in the liquid state with a high fluidity. It quickly becomes very viscous, in the early stage of solidification, and later completely solidifies (Gafur et al, 2003). For the numerical analysis of heat transfer problem, the appropriate set of equations were determined that described the heat transfer behaviour in the cast metal (Hearn, 1992). With the boundary conditions, initial conditions, and thermo-physical properties of the materials being known, it is possible to obtain the temperature and variation of the whole casting system (Ozisik, 1985) and (Liu et al (1993). Finite number at discrete points (Adams and Rogers,(1973), Shampire,(1994) and Bayazitoglu and Ozisik,(1988)) within the cast specimen was employed as the numerical method of solution. This method provides the temperature at a discrete number of points in the cast region. In the numerical method, the cast region is defined and divided into discrete number of points. As temperature difference is imposed in the system, heat flows from the high-temperature region to the low-temperature region as shown in figure 1.

To determine the temperature distribution, energy conservation equations were used for each of the nodal points of the unknown temperature at the interfaces and the cast regions((Incropera and Dewitt, 1985) and (Janna, 1988)). Temperatures were monitored at distance 2mm into the cast metal, represented by grid point M, and at the steel mould/cast metal interface (see figure 4). By using measured temperatures in both the casting and the steel mould, together with the numerical solutions of the solidification problem, heat transfer coefficients were determined based on Beck (1970)solution of the inverse heat conduction problem. The estimation of the surface heat transfer coefficients or heat flux density utilizing a measured temperature history inside a heat-conducting solid is called the inverse heat conduction problem (Cho and Hong 1996). This problem becomes non-linear, as the thermal properties (thermal conductivity, specific heat) are temperature dependent.

EXPERIMENTAL PROCEDURE

Chromel-Alumel thermocouples TC2, TC3, TC4, TC5 and TC6 were positioned on the sides of the cylindrical steel container, while TC1 and TC7 were positioned in the cast aluminium metal in the cylindrical and bottom flat surfaces respectively as shown in figure 4below.

Thermocouples of chromel-alumel type, 3mm in diameter were used to determine the solidifying temperatures of the cast molten metal and heating temperatures of the steel mould at the various positions in the cylindrical steel container of figure 4. The solidifying temperatures at both the cylindrical and flat bottom surfaces of the cast molten aluminium metal were monitored at a position 2mm (from the surface of the steel mould –cast aluminium metal interface) into the cast molten aluminium metal.

At the steel mould wall in the cylindrical surface, thermocouples were positioned at $X2 = 4$mm, $X3 = 8$mm, $X4 = 12$mm, $X5 = 16$mm and $X6 = 20$mm measured from the cast aluminium metal / steel mould interface to monitor the heating temperatures at these positions of the steel mould wall as shown in fig. 4. From the temperatures versus time curves obtained for each position in the steel mould, the interface heating temperature versus time curve at the cast aluminium metal / steel mould, for position when $X = 0$ was obtained by using the polynomial curve fitting method.

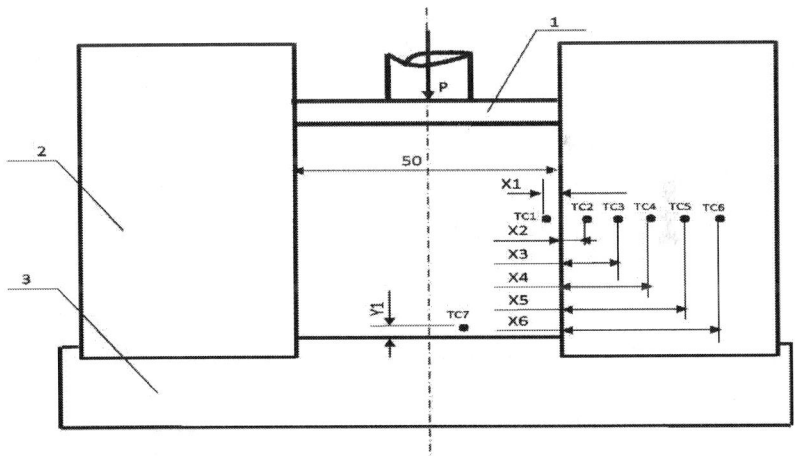

1-upper punch, 2-cylindrical steel mould, 3-lower die ($X1 = 2$, $X2 = 4$, $X3 = 8$, $X4 = 12$, $X5 = 16$, $X6 = 20$, $Y1 = 2$ (mm))

Figure 4: Schematic diagram of squeeze casting test rig.

This was done by selecting a particular time of heating of steel mould, say $t = 10$sec. and drawing vertical lines cutting across the heating temperatures versus time curves at various thermocouples' distances within the steel mould. At the point of intersection with each curve, the value of temperature was read against distance X, for the chosen time, t = 10sec. The value of interface steel mould / cast temperature at time say, t = 10sec.was determined at the steel mould / cast metal interface by substituting the value of X = 0 in the polynomial curve fitting equation(70) obtained from the values of temperatures at various distances in the steel mould at a chosen time, t = 10sec.

$$T_{X0} = 0.0031X^4 - 0.168X^3 + 3.263X^2 - 22.812X + 117.8 \qquad (70)$$

The temperature obtained by this method corresponds to the interface steel mould / cast metal temperature at a distance X = 0 for the chosen time t = 10sec. If this procedure is repeated for a number of time increments, the temperatures obtained with corresponding times represent the temperature at X=0, for such time increments. The graph of extrapolated temperatures versus time is drawn for position when X = 0 to represent the heating temperatures versus time curve at the steel mould / cast aluminium metal interface is shown in figure 5.

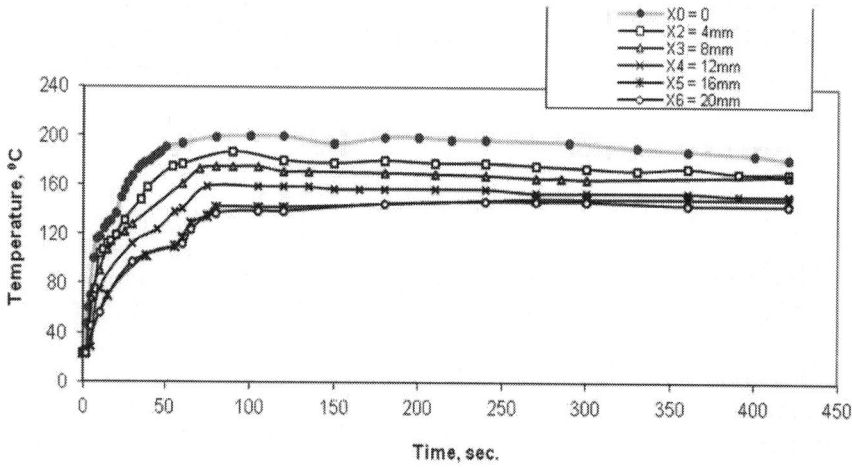

Figure 5: Effect of distance on the heating temperatures of steel mould (extrapolated heating curve at the cast specimen/steel mould interface i.e. X=0).

INTERFACE HEAT TRANSFER COEFFICIENTS DETERMINATION

Extrapolated temperature versus time curve of figure 5 for position when X = 0 (i.e. cast aluminium metal / steel mould interface) was used to determine the heat transfer coefficients of solidifying molten aluminium metal. It was used to determine the interface heat transfer coefficients in the cast aluminium metal / steel mould for no pressure and with pressure applications at both the cylindrical and bottom flat surfaces of the steel mould.

The interface heat transfer coefficients between the steel mould and cast aluminium metal at the cylindrical and bottom flat surfaces were determined from the extrapolated experimental heating temperature versus time curve obtained for position X = 0 and aluminium cast solidification temperature versus time curves obtained for the cylindrical and bottom flat surfaces, using equations (66) and (67).

The interface heat transfer coefficients were determined also numerically by the inverse method using the Finite Difference Method (FDM) and the obtained results were compared with the experimentally derived values.

DISCUSSIONS OF RESULTS

Temperature-Time Curves

Figure 6 shows typical temperature versus time curves for solidifying molten aluminium metaland steel mould respectively without the application of pressure on solidifying metal.

This figure shows the comparison of the numerical method usually applied by Cho and Hong (1996) to determine interface steel mould / cast metal temperature versus time curve with the extrapolated experimental method of this present work. The heating curve, as obtained through extrapolations of polynomial curves fitting equations and numerical methods are in close agreement and the deviations from the values obtained numerically varied from between 1.26- 19.31%.

Typical result obtained under pressure is also shown in figure 7, indicating the solidification and heating curves generated for solidifying molten aluminium metal and steel mould which follow the same patterns to the curves in figure 6.

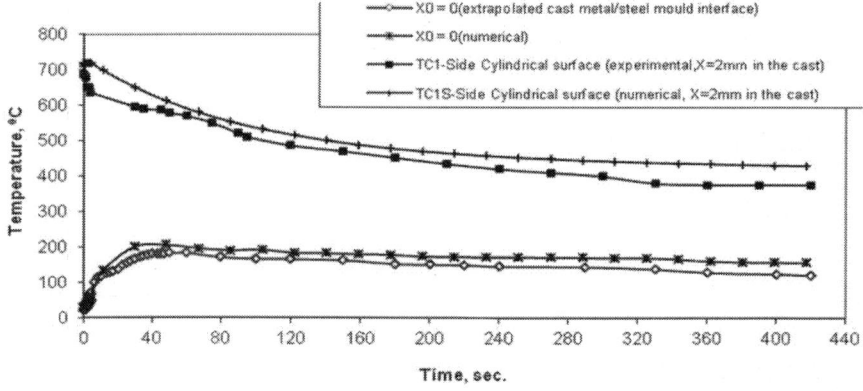

Figure 6: Comparison of experimental measured temperatures with numerical values of aluminium metal without pressure application (P = 0).

With the application of pressure, the peak temperatures recorded are about the same 649°C and 648°C for a pressure of 85.86 MPa at the bottom flat and cylindrical surfaces of the steel mould respectively (see figure 7). The peak temperature (649°C) obtained at the bottom flat surface of the steel mould under applied pressure is found to be higher than that temperature (607°C) without pressure application. This effect may be associated to additional internal heat generated, resulting to higher temperature during pressure application on the solidifying molten aluminium.

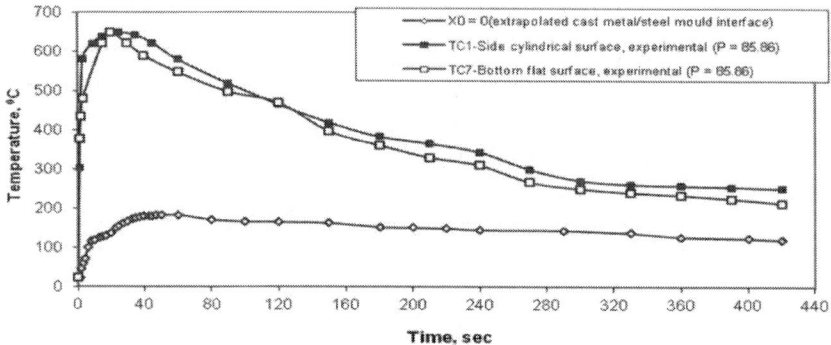

Figure 7: Effect of pressure on the experimental measured temperatures of solidification of aluminium metal (P = 85.86MPa) for side and bottom mould's surface.

Interface Heat Transfer Coefficients with Time

From the temperature with time curves of figure 6, the heat transfer coefficients for both cylindrical and bottom flat surfaces were determined for both numerical and calculated values and shown in figure 8. The maximum heat transfer coefficients of 2927.92 W/m²K and 2975.14 W/m²K are obtained at the cylindrical and bottom flat surfaces respectively for no pressure application, which is close to 2900 W/m²K as obtained for pure aluminium by Kim and Lee (1997). The values of heat transfer coefficients decrease rapidly for both the cylindrical and bottom flat surfaces to a level of 866.70 W/m²K and 969.50 W/m²K respectively in 90 seconds. These values further decrease to 361.80 W/m²K and 478.80 W/m²K at these surfaces in another 150 seconds and further decrease then becomes not so noticeable.

From figure 8, the peak values of interface heat transfer coefficients are 2927.92 W/m²K and 2956.73 W/m²K as obtained by experimental and numerical determinations respectively at the cylindrical surface for no pressure application. For times within 40 seconds to 120 seconds the values of the interfacial heat transfer coefficients obtained numerically and experimentally are found to show higher values of about 19.83 % for numerical results to experimental results.

With the application of pressure on the solidifying aluminium metal, the heat transfer coefficients reach maximum values of

3085.34 W/m²K and 3351.08 W/m²K in the cylindrical and bottom flat surfaces respectively (see figure 9). These values also decrease to 847.80 W/m²K and 783.63 W/m²K in 240 seconds in the cylindrical and bottom flat surfaces respectively, while further decrease with time of solidification is no longer pronounced.

Interface Heat Transfer Coefficients with Solidification Temperatures

Figures 10 and 12 show the calculated experimental interface heat transfer coefficients for solidifying molten aluminium metal as a function of solidification temperatures of the solidifying molten aluminium metal. Figure 10 shows the variation of heat transfer coefficient with solidification temperatures of aluminium at the cylindrical surface, while figure 12 is the interface heat transfer coefficients with solidifying temperature at the bottom flat surface with and without the application of pressure on the solidifying metal. From the two graphs, the maximum interface heat transfer coefficients obtained without pressure and with pressure application in the bottom flat surface of the steel mould are 2975.14 W/m²K and 3351.08 W/m²K respectively.

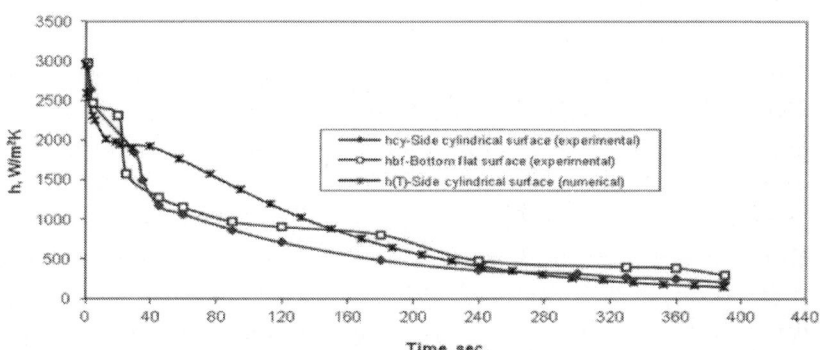

Figure 8: Comparison of numerical values of heat transfer coefficients with calculated experimental values (P = 0).

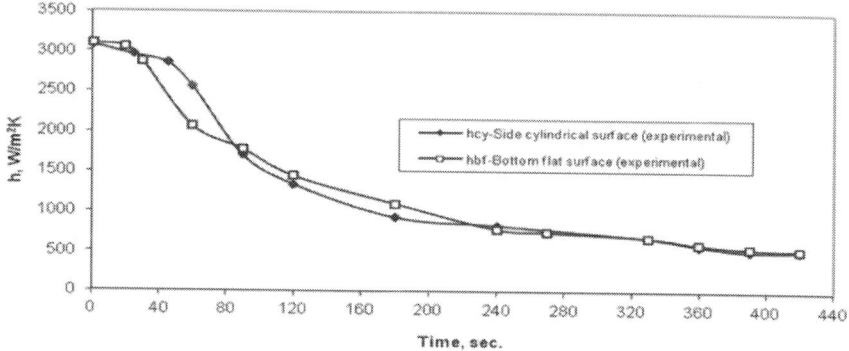

Figure 9: Effect of time of solidification of aluminium metal on heat transfer coefficients with pressure application (P = 85.86MPa) at side and bottom surfaces of steel mould.

Figure 11 shows the numerical values of the variation of interface heat transfer coefficients with the solidification temperature of aluminium metal at the cylindrical surface with the application of pressure. The maximum value of heat transfer coefficient of 3397.29 W/m^{2}K at applied pressure of 85.86Mpa as compared to 3351.08 W/m^{2}K obtained through the experimental procedure.

At solidifying temperature above 600°C, a sharp reduction in the interface heat transfer coefficients is noticed at both surfaces as is shown on figures 10, 11 and 12.

For temperatures below 500°C, the interface heat transfer coefficients for both no pressure and pressure applications are close in values. This shows that at temperature below 500°C, the effect of applied pressure is no longer significant on the interface heat transfer coefficient values. The drop in temperature results in solidification of the molten aluminium, which in turn leads to a drop in the heat transfer coefficient values. The effect of applied pressure on the heat transfer coefficients of aluminium becomes more pronounced at solidifying temperatures above 500°C which was also reported by Cho and Hong (1996). Below this temperature, the effect of applied pressure on interface heat transfer coefficient values becomes less pronounced.

Therefore, from figures 10, 11 and 12, it is observed that the effect of applied pressure becomes more significant at temperature close to the liquidus temperature of aluminium as measured along the bottom flat surface of the steel mould (see figure 12). The maximum value of 3397.29 W/m²K is obtained for pressure level of 85.86MPa as compared to 2975.14 W/m²K for no pressure at the bottom flat surface of the steel mould.

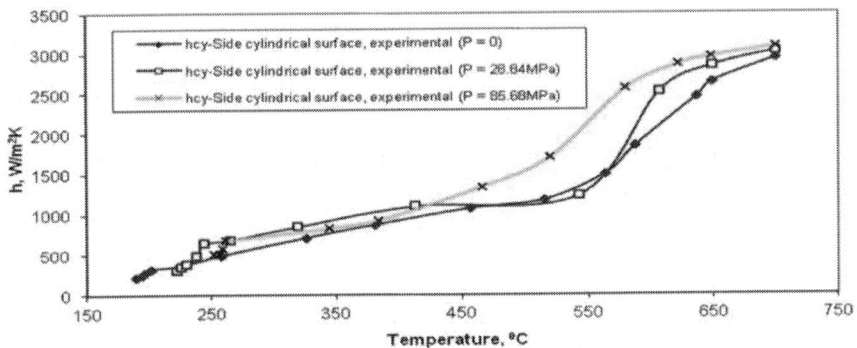

Figure 10: Typical effects of pressure applications on heat transfer coefficients with solidifying temperature at the side cylindrical mould surface by experimental method.

Peak Interface Heat Transfer Coefficients with Applied Pressures

Figure 13 shows the variation of peak values of interface heat transfer coefficients with and without pressure applications. Higher experimental values of heat transfer coefficients are obtained at the bottom flat surface than at the cylindrical surface of the steel mould (see figure 13). This can be associated to greater effect of pressure application experienced at the bottom flat surface than at the side cylindrical surface, thus leading possibly to greater additional internal heat energy generated, and hence obtained higher values of heat transfer coefficients. The results of numerical determination of heat transfer coefficients as in figure 13 shows higher values as compared to the values obtained by experimental method. At applied pressure of 85.86MPa, the obtained heat transfer coefficients are 3397.29 W/m²K and 3351.08 W/m²K by numerical and experimental procedures respectively. From the curves of heat transfer coefficients obtained with temperatures of figures 10, 11 and 12, three distinct portions are noticed. These portions are easily differentiated by the aluminium solidification temperature. These temperatures are below 500°C, solidus phase, 500 to 660°C, liquidus-solidus phase, and above 660°C liquidus phase of solidification of molten aluminium.

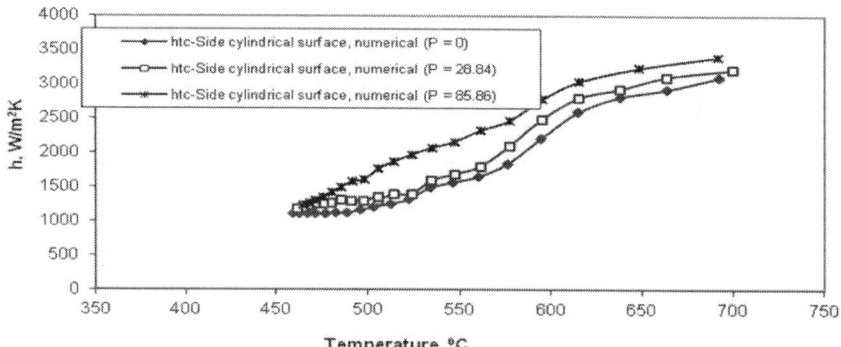

Figure 11: Typical effects of pressure application on heat transfer coefficients with solidifying temperature of the side cylindrical mould surface by numerical method.

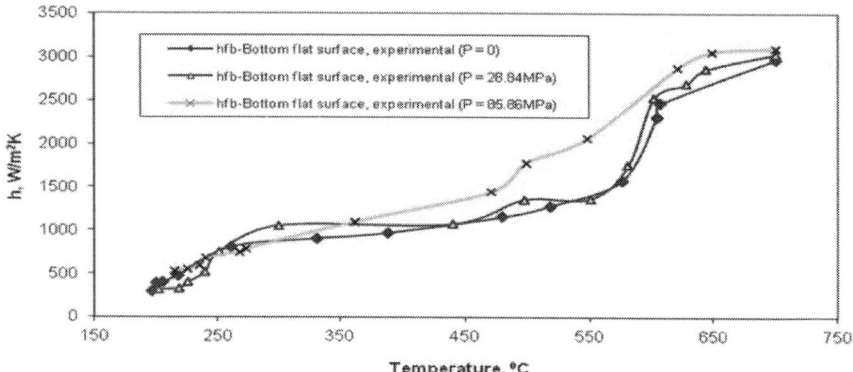

Figure 12: Typical effects of pressure applications on the heat transfer coefficients with solidifying temperature at the bottom flat mould surface.

The liquidus-solidus phase, which occurs over a solidification temperature range of 500℃ to 660℃, instead of a constant solidification temperature of 660℃ can be attributed to the presence of impurities such as silicon, magnesium and manganese in the commercially pure aluminium a fact supported by Higgins (1983).

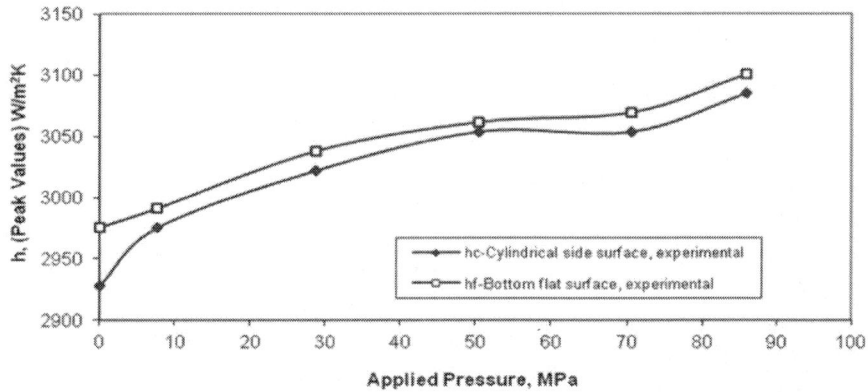

Figure 13: Effect of pressure on the peak values of heat transfer coefficients of aluminium metal at liquidus stage at side and bottom flat moulds' surfaces.

The empirical equations for each of the distinct phase changes as a function of applied pressures and solidification temperatures are determined for both the experimental and numerical methods. The empirical equations obtained are for mean average values of heat transfer coefficients based on the experimental method at the cylindrical and flat bottom surfaces. These are equations (71-73):

Temperatures below 500℃ (solidus phase),

Temperatures below 500°C (solidus phase),

$$h_{S(exp)} = 3.081T + 1.303P - 232.942 \tag{71}$$

with coefficient of correlation r = 0.9545.

Temperatures between 500°C and 660°C (liquidus-solidus phase),

$$h_{LS(exp)} = 10.420T + 5.641P - 4176.022 \tag{72}$$

with coefficient of correlation r = 0.9884.

Temperatures above 660°C (super heat, liquidus phase),

$$h_{L(exp)} = 2.769T + 2.518P + 988.921 \tag{73}$$

with coefficient of correlation r = 0.7825.

The empirical equations (74-76) obtained through the numerical methods are from the results of the computer simulations of heat transfer coefficients at the cylindrical cast metal / steel mould interface by the application of various applied pressures. These empirical equations are:

Temperatures below 500°C (solidus phase),

$$h_{S(num)} = 3.849T + 3.643P - 700.427 \tag{74}$$

with coefficient of correlation r = 0.969

Temperatures between 500°C and 660°C (liquidus-solidus phase),

$$h_{LS(num)} = 9.027T + 3.414P - 3000.625 \tag{75}$$

with coefficient of correlation r = 0.964

Temperatures above 660°C (super heat, liquidus phase),

$$h_{L(num)} = 2.489T + 2.787P + 1342.19 \tag{76}$$

with coefficient of correlation r = 0.772

Die Heating Effect

Figure 14 is the effect of die pre-heat temperatures on the values of heat transfer coefficients of aluminium metal without the application of pressure. From the figure, the heat transfer coefficients become lower with increase in die pre-heat temperatures. At die temperature of 95ºC, the heat transfer coefficient is 3185.34 W/m²K and drop to a value of 2476.73 W/m²K at die temperature of 300ºC. For all the die temperatures, there is a fall in the heat transfer coefficient's values as solidification temperature decreases.

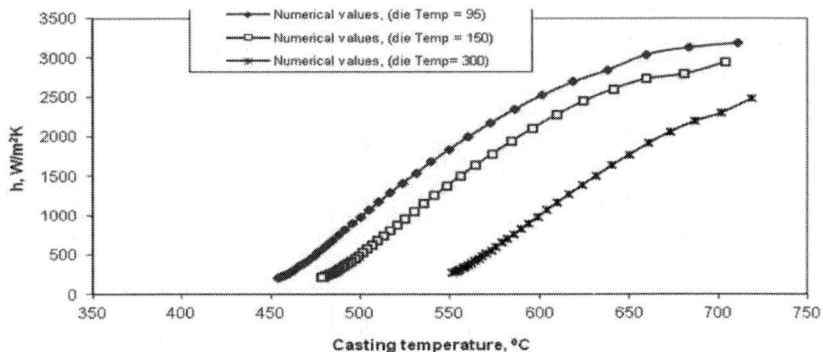

Figure 14: Typical effect of die temperature on heat transfer coefficients without the application of pressure (P=0).

Comparison of Heat Transfer Coefficient with Semi-Empirical Method

The values of heat transfer coefficients determined using experimental, heat differential equations (numerical), and methods of semi-empirical equations are shown in figure 15 under a pressure application of 85.86MPa. From this graph, the peak values of interface heat transfer coefficient are 3358.19 W/m²K and 3198.79 W/m²K as obtained by heat differential and method of semi-empirical equations respectively for a pressure application of 85.86MPa. The heat transfer coefficients' values for the three methods drop with time and are found to be 1708.03, 1976.81 and

1838.72 W/m²K in 100seconds for experimental, differential and methods of semi-empirical equations respectively.

With die temperature of TM=150⁰C, the peak heat transfer coefficients of 3088.99 W/m²K, 3249.84 W/m²K and 2982.60 W/m²K are obtained for experimental, heat differential and method of semi-empirical equations as shown on figure 16 following the same pattern as in figure 15.

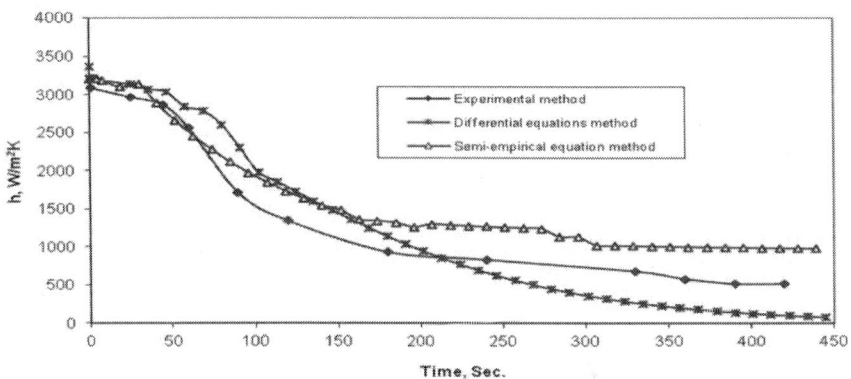

Figure 15: Typical comparison of numerical values of interface heat transfer coefficients with experimental and empirical values with pressure application (P=85.86MPa, TM =300C).

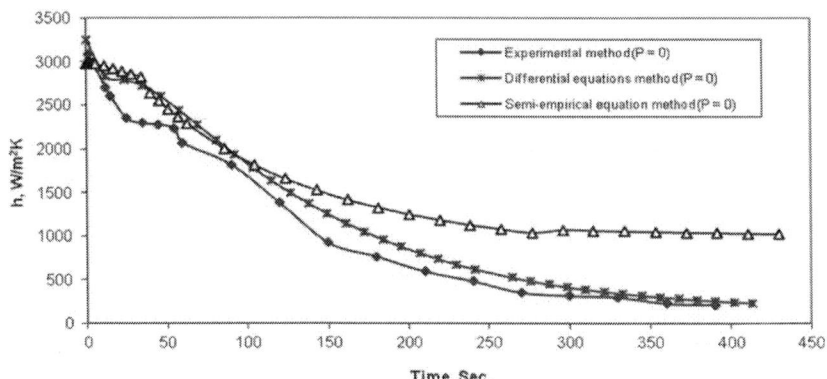

Figure 16: Typical comparison of numerical values of interface heat transfer coefficients with experimental and empirical values with die heating (TM=150°C).

CONCLUSIONS

The following conclusions can be made from the present investigation:

The graph of temperature against time curves obtained by extrapolating to steel mould / cast metal interface by polynomial curve fitting to heating temperatures graphs with times at various steel mould locations are found to agree in values to the usual numerical methods obtained by previous authors. The interface heat transfer coefficients obtained by the numerical and experimental methods without the application of pressure are found to have values close to that of the numerical methods. The values of the numerical methods were higher by about 19.83%.

Effect of pressure application on the solidifying molten aluminium is more pronounced at casting temperatures above 500^0C of the cast aluminium specimen on the values of the interface heat transfer coefficients obtained. Interface heat transfer coefficients are found to decrease with increase in solidification time in both the cylindrical and bottom flat surfaces of the steel mould and thereafter remain fairly constant at temperature below 500^0C.

Values of experimental peak heat transfer coefficients at the bottom flat surface are found to be higher with pressure application on the solidifying aluminium metal than at the cylindrical surface.

The empirical equations, relating the values of interface heat transfer coefficients with the applied pressures and solidification temperatures at three distinct stages of solidifying molten aluminium are determined and can be applied to determine the heat transfer coefficients.

The values of heat transfer coefficients obtained by heat differential equations incorporating internal heat energy and methods of semi-empirical equations are very close in values. The values as obtained by semi-empirical equations were higher by about 1.7% within the first 5 seconds of solidification.

The semi-empirical equations generated are flexible and could be used to predict the casting temperatures of other metals if the heat transfer coefficient values at the three phase changes are known.

REFERENCES

1. J. Adams, Alan, David. F. Rogers, 1973Computer Aided Heat Transfer Analysis", McGraw-Hill Publishing Company, Tokyo.
2. Yildiz. Bayazitoglu, M. Ozisik, Necati, 1988Experimental Determination of Heat Transfer Coefficients During Squeeze Casting of AluminiumMcGraw-Hill Book Company, New York.
3. J. V. Beck, 1970Experimental Determination of Heat Transfer Coefficients During Squeeze Casting of AluminiumInt. J. Heat Mass Transfer, 13703716
4. W. Bolton, 1989Experimental Determination of Heat Transfer Coefficients During Squeeze Casting of AluminiumButterworths-Heinemann Limited, UK.
5. David. J. Browne, D. O'Mahoney, 2001Interface Heat Transfer in Investment casting of Aluminium", Metallurgical and Materials Transactions A, Dec. 32A30553063
6. William. D. Callister, Jr , 1997Experimental Determination of Heat Transfer Coefficients During Squeeze Casting of Aluminiumth Edition, John Wiley and Sons Inc.
7. Himadri. Chattopadhyay, 2007Experimental Determination of Heat Transfer Coefficients During Squeeze Casting of AluminiumJ. Materials Processing Technology, 186, 174178
8. S. Cho, C. P. Hong, 1996Experimental Determination of Heat Transfer Coefficients During Squeeze Casting of AluminiumInt. J. Cast Metals Res., 9227232
9. Das, S. Chatterjee, 1981Squeeze Casting of an Aluminium Alloy Containing Small Amount of Silicon Carbide Whiskers", The Metallurgist and Materials Technologist, 137142
10. R. Elliott, 1988Experimental Determination of Heat Transfer Coefficients During Squeeze Casting of AluminiumButterworths, London, U.K.
11. M. A. Gafur, Haque. Nasrul, K. , Narayan Prabhu, (2003Effects of chill Thickness and Superheat on Casting/Chill Interfacial Heat Transfer During Solidification of Commercially Pure Aluminium", J Materials Processing Technology, 133, 257265
12. E. J. Hearn, (1992), "Mechanics of Materials", 2nd Edition, Pergamon Press, UK.

13. Raymon. A. Higgins, 1983Engineering Metallurgy Part I: Experimental Determination of Heat Transfer Coefficients During Squeeze Casting of Aluminiumth Edition, ELBS with Edward Arnold, UK.

14. William. F. Hosford, Robert. M. Caddell, 1993Experimental Determination of Heat Transfer Coefficients During Squeeze Casting of Aluminiumnd edition, PTR Prentice-Hall Englewood, NJ.

15. H. Hu, A. Yu, 2002Experimental Determination of Heat Transfer Coefficients During Squeeze Casting of AluminiumModelling Simul. Mater Sci. Eng., 10111

16. Frank. P. Incropera, David. P. Dewitt, 1985Fundamental of Heat and Mass Transfer", 3rd Edition, John Wiley and Sons, NY.

17. William. S. Janna, 1988Experimental Determination of Heat Transfer Coefficients During Squeeze Casting of AluminiumSI Edition, Van Nostrand Reinhold (International0, U.K.

18. T. G. Kim, Z. H. Lee, 1997Experimental Determination of Heat Transfer Coefficients During Squeeze Casting of AluminiumInt J, heat mass transfer, 40(15), 35133525

19. P. A. Kobryn, S. L. Semiatin, 2000Experimental Determination of Heat Transfer Coefficients During Squeeze Casting of AluminiumMetallurgical and materials Transactions, August, 32B685695

20. Liu, T. E. Voth, T. L. Bergman, 1993Experimental Determination of Heat Transfer Coefficients During Squeeze Casting of AluminiumInt. J. Heat Mass Transfer, 362441442

21. Maleki, B. Niroumand, A. Shafyei, 2006Experimental Determination of Heat Transfer Coefficients During Squeeze Casting of AluminiumMaterials Science and Engineering A, 428 135140

22. M. A. Martorano, J. D. T. Capocchi, 2000Heat Transfer Coefficient at the Metal-Mold Interface in the Unidirectional Solidification of Cu-8%Sn alloys", Intl J. Heat Mass Transfer, 43, 25412552

23. M. Ozisik, Necati, 1985Heat Transfer:- A Basic Approach", McGraw-Hill Publishing, Company U.K.

24. D. A. Potter, K. E. Easterling, 1993Phase Transformations in Metals", 2nd Edition, Chapman & Hall, London.

25. R. E. Reed-Hill, R. Abbaschian, 1973Physical Metallurgy Principles", 3rd Edition, PWS-KENT Publishing Company, Boston.

26. A. Santos, J. M. V. Quaresma, A. Garcia, 2001Determination of Transient Heat Transfer Coefficients in Chill Mold Castings", Journal of Alloys and Compounds, 139, 174186

27. A. Santos, A. Garcia, C. R. Frick, J. A. Spim, 2004Evaluation of heat transfer coefficients along the secondary cooling zones in the continuous casting of steel billets, Inverse problems, Design and Optimization symposium", Rio de Janeiro, 18

28. Lawrence. F. Shampire, 1994Numerical Solution of Ordinary Differential Equations", Chapman & Hall, New York.

29. Frank. M. White,. , 1991Heat and Mass Transfer", Addison Wesley Publishing Company, Massachusetts.

CITATION

Jacob O. Aweda and Michael B. Adeyemi (2012). Experimental Determination of Heat Transfer Coefficients During Squeeze Casting of Aluminium, An Overview of Heat Transfer Phenomena, Dr M. Salim Newaz Kazi (Ed.), ISBN: 978-953-51-0827-6, InTech, DOI: 10.5772/52038.

CHAPTER 5

Heat Transfer Analysis of Mhd Thin Film Flow of An Unsteady Second Grade Fluid Past A Vertical Oscillating Belt

Taza Gul1 , Saeed Islam1 , Rehan Ali Shah², Ilyas Khan³, Asma Khalid⁴ Sharidan Shafie⁴

¹Department of Mathematics, Abdul Wali Khan University, Mardan Khyber Pakhtunkhwa, Pakistan,

²Department of Mathematics, University Of Engineering And Technology, Peshawar Khyber Pakhtunkhwa, Pakistan,

³Department of Basic Sciences, College of Engineering Majmaah University, Majmaah, Saudi Arabia,

⁴Department of Mathematical Sciences, Faculty of Science, University Teknology Malaysia, UTM Johor Bahru, Johor, Malaysia

INTRODUCTION

Thin-film flow is significant regarding broad class of physical applications and attracts the attention of physicists, engineers and chemists. In the field of chemical engineering, thin film layers are functioning to design efficient and gainful development units such as thin-film reactors, evaporators, condensers, distillation columns and heat exchangers. The enormous benefit of thin film layers is related to their tiny thickness which, in turn, results in large heat- and mass-transfer areas per unit volume. Further, thin fluid layers have been executed in circumstances where a film of fluid layers is over a solid surface such as in different coating processes [1]. At the micron scale, thin layer is of particular importance, specified by a large scale of microfluidic devices, as evaluated in the work of Stone et al. [2] and Squires and Quake[3].

In physical, chemical and biological sciences, thin film flows have been used in micro-channel heat sinks to provide cooling for nanotechnologies. In environmental and geophysical engineering, thin film flows have been related with geological problems such as lava, debris flows and mudslides [4], [5];

Keeping in view the rich applications of non-Newtonian fluids in engineering and industry, such fluids have been widely studied. Ample research has been carried out in this field. Considerable efforts have been made to study non-Newtonian fluids through analytical and numerical treatment.

One of the well-known model amongst non-Newtonian fluids is the class of second grade fluids which has its constitutive equations based on strong theoretical foundations. Some development and relevant work on this topic is the wire coating in a straight annular die for unsteady second grade fluid discussed by Rehan et al. in [6].

They modeled the unsteady second grade fluid flow between wire and die with one oscillating boundary and the other stationary in the form of partial differential equation. Similar results can also be found in [7], [9]. On the other hand, Samiulhaq et al. [10] investigated unsteady free convection flow of a second grade fluid. They have compared the influence of ramped temperature and isothermal temperature on the velocity field and skin friction through different cases in the presence of magnetic field as well as porosity. Ali et al. [11] studied the closed form solutions for unsteady second grade fluid near vertical oscillating plate. They have shown the effect of various physical parameters on the velocity and temperature fields.

The physical importance of thin film has been researched and discussed by several authors. For examples, thin film flow of a power law model liquid falling an inclined plate was discussed by Miladinova et al. [12], wherein they observed that saturation of non-linear interaction occurred in a finite amplitude permanent wave. Alam et al. [13] investigated the thin-film flow of Johnson-Segalman fluids for lifting and drainage problems. They observed the effect of various parameters on the lift and drainage velocity profiles. To solve real world problems, several approximate techniques have been used in mathematics, fluid mechanics and engineering sciences. Some of the common methods are, HAM and OHAM [14], [15]. Application of optimal Homotopy asymptotic method for solving non-linear equations arising in heat transfer was investigated by Marinca and Herisanu [16]. They have also discussed an optimal Homotopy asymptotic method applied to steady flow of a fourth-grade fluid past a porous plate [17]. These methods deal with the nonlinear problems effectively. Mabood et al. [18] discussed OHAM solution of viscoelastic fluid in axisymmetric heated channels. They have shown that the results of OHAM are comparatively better than other methods' results. Some development in this direction is discussed in [19]–[27]. Taza Gul et al. [28] investigated effects of MHD on thin film flow of third grade fluids for lifting and drainage problems under the action of heat dependent viscosity. The effects of various parameters on the lift and drainage velocity profiles are also studied.

The main objective of this work is to study the effects of oscillation into a MHD thin film flow of an unsteady second grade fluid on a vertical oscillating belt using ADM and OHAM. In 1992, Adomian [29], [30] introduced the ADM for the approximate solutions for linear and non linear problems. Wazwaz [31], [32] used ADM for the reliable treatment of Bratu-type and Rmden-Fowler equations. In a comparative study, Taza Gul et al. [33] used ADM and OHAM for solution of thin film flow of a third grade fluid on a vertical belt with slip boundary conditions.

The convergence of the decomposition series was cautiously examined by several researchers to verify the fast convergence of the resulting series. Cherruault examined the convergence of Adomian's method in [34]. Cherruault and Adomian presented a new proof of convergence of the method in [35].

Basic Equations

The constitutive equations governing the problem (equation of continuity, momentum and energy) under the influence of externally imposed transverse magnetic field are:

$$\nabla.\mathbf{u}=0 \tag{1}$$

$$\rho\frac{D\mathbf{u}}{Dt}=\nabla.\mathbf{T}+\rho g+\mathbf{J}\times\mathbf{B}, \tag{2}$$

$$\rho\,c_p\frac{D\Theta}{Dt}=k\,\nabla^2\,\Theta+tr(\mathbf{T}.\mathbf{L}), \tag{3}$$

where ρ, is the constant density, g denotes gravity, u is velocity vector of the fluid, Θ defines temperature, k is the thermal conductivity, C_P is specific heat,

$$\mathbf{L}=\nabla\mathbf{u},\frac{D}{Dt}=\frac{\partial}{\partial t}+(\mathbf{u}.\nabla)$$

denotes material time derivative, and T is the Cauchy stress tensor.

One of the body force term corresponding to MHD flow is the Lorentz force J x B. Where B is the total magnetic field and J is the current density. By using Ohm's law, the current density is given as

$$\mathbf{J}=\sigma(\mathbf{E}+\mathbf{V}+\mathbf{B})$$

where σ is electrical conductivity of the fluid, E is the electric field, V is the velocity vector field, B = B_0 + b_1 with B_0 is the imposed magnetic field and b_1 is the induced magnetic field. The current density J with the assumptions E = 0, B = 0 and, B = B_0 = (0, B_0, 0)where B_0 is the strength of applied magnetic field B_0, modifies to J = σ (V × B_0). Finally the Lorentz force becomes

$$\mathbf{J}\times\mathbf{B}=\left[0,\sigma\,\mathbf{B}_0^2\,u(x,t),0\right], \tag{4}$$

Cauchy stress tensor **T** is given by

$$\mathbf{T}=-p\mathbf{I}+\mathbf{S}, \tag{5}$$

where $-p\mathbf{I}$ denotes spherical stress and shear stress **S**, is defined as

$$S = \mu A_1 + \alpha_1 A_2 + \alpha_2 A_1^2, \tag{6}$$

α_1 and α_2 are the material constants and A_1, A_2 are the kinematical tensors given by

$$A_1 = (\nabla u) + (\nabla u)^T,$$

$$A_n = \frac{D A_{n-1}}{Dt} + A_{n-1} (\nabla u) + (\nabla u)^T A_{n-1}, n \geq 2 \tag{7}$$

Formulation Of The Lift Problem

Consider, a wide flat belt moves vertically at time $t = 0^+$, the belt is oscillated and translated with constant speed U through a large bath of second grade liquid. The belt carries a layer of liquid of constant thickness δ. Coordinate system is chosen for analysis in which the y-axis is taken parallel to the belt and x-axis is perpendicular to the belt. Uniform magnetic field is applied transversely to the belt. It has been assumed that the flow is unsteady and laminar after a small distance above the liquid surface layer.

Velocity and temperature fields are defined as:

$$u = (0, u(x,t), 0), \Theta = \Theta(x,t) \tag{8}$$

Oscillating boundary conditions are:

$$u(0,t) = U(1 + \xi Cos\omega t), \frac{\partial u(\delta,t)}{\partial x} = 0, \tag{9}$$

$$\Theta(0,t) = \Theta_0, \Theta(\delta,t) = \Theta_1, \tag{10}$$

Here ξ is used as amplitude in [6] and [9]. ω is used as frequency of the oscillating belt.

Inserting the velocity field from Eq.(8) in continuity Eq.(1) and in momentum Eqs.(2) and (4), the continuity Eq.(1) is satisfied identically and momentum Eqs. (2) and (4) are reduced to the following components of stress tensor as:

$$T_{xx} = -P + (2\alpha_1 + \alpha_2)\left(\frac{\partial u}{\partial x}\right)^2,$$

(11)

$$T_{xy} = \mu\frac{\partial u}{\partial x} + \alpha_1\frac{\partial}{\partial t}\left(\frac{\partial u}{\partial x}\right),$$

(12)

$$T_{yy} = -P + \alpha_2\left(\frac{\partial u}{\partial x}\right)^2,$$

(13)

$$T_{zz} = -P,$$ (14)

$$T_{xz} = T_{yz} = 0,$$ (15)

making use of Eqs. (11–15) in Eq.(2,3), the momentum and energy Eqs. (2,3) are reduced to,

$$\rho\frac{\partial u}{\partial t} = -\frac{\partial p}{\partial y} + \mu\frac{\partial^2 u}{\partial x^2} + \alpha_1\frac{\partial}{\partial t}\left(\frac{\partial^2 u}{\partial x^2}\right) - \rho g - \sigma B_0^2 u,$$

(16)

$$\rho c_p\left(\frac{\partial \Theta}{\partial t}\right) = k\frac{\partial^2 \Theta}{\partial x^2} + \mu\left(\frac{\partial u}{\partial x}\right)^2 + \alpha_1\left(\frac{\partial u}{\partial x}\right)\frac{\partial}{\partial t}\left(\frac{\partial u}{\partial x}\right),$$

(17)

Introducing the following non-dimensional variables

$$\breve{u} = \frac{\mathbf{u}}{U}, \breve{x} = \frac{x}{\delta}, \breve{t} = \frac{\mu t}{\rho \delta^2}, \breve{\Theta} = \frac{\Theta - \Theta_0}{\Theta_1 - \Theta_0}, B_r = \frac{\mu U^2}{k(\Theta_1 - \Theta_0)},$$

$$M = \frac{\sigma B_0^2 \delta^2}{\mu_0}, P_r = \frac{\mu c_p}{k}, \breve{\omega} = \frac{\omega \delta^2 \rho}{\mu},$$

$$S_t = \frac{\delta^2 \rho g}{\mu U}, \alpha = \frac{\alpha_1}{\rho \delta^2},$$

$$(18)$$

where ω is the frequency parameter, α is non-Newtonian effect, M is magnetic parameter, t is time parameter, B_r is Brinkman number, S_t is Stock's number and P_r is the Prandtl number.

On inserting the above dimensionless variables in Eqs. (16, 17), when , $\delta P / \delta y$ the momentum and energy equations become,

$$\frac{\partial u}{\partial t} = \frac{\partial^2 u}{\partial x^2} + \alpha \frac{\partial}{\partial t}\left(\frac{\partial^2 u}{\partial x^2}\right) - S_t - Mu,$$

$$(19)$$

$$P_r \left(\frac{\partial \Theta}{\partial t}\right) = \frac{\partial^2 \Theta}{\partial x^2} + B_r \left[\left(\frac{\partial u}{\partial x}\right)^2 + \alpha \left(\frac{\partial u}{\partial x}\right)\left(\frac{\partial^2 u}{\partial t \partial x}\right)\right],$$

$$(20)$$

From Eqs. (9, 10), the non-dimensional boundary conditions are:

$$u(0,t) = 1 + \xi Cos\omega t, \frac{\partial u(1,t)}{\partial x} = 0,$$

$$(21)$$

$$\Theta(0,t) = 0, \Theta(1,t) = 1,$$

$$(22)$$

Analysis of Adomain Decomposition Method

The Adomian Decomposition Method (ADM) is used to decompose the unknown function $u(x, y)$ into a sum of an infinite number of components

defined by the decomposition series.

$$u(x,t) = \sum_{n=0}^{\infty} u_n(x,t),$$

$$(23)$$

The decomposition method is used to find the components $u_0(x,t), u_1(x,t), u_2(x,t), \cdots$ separately. The determination of these components can be obtained through simple integrals.

To give a clear overview of ADM, we consider the linear partial differential equation in an operator form as

$$L_t u(x,t) + L_x u(x,t) + Ru(x,t) + Nu(x,t) = g(x,t), \tag{24}$$

$$L_x u(x,t) = g(x,t) - L_t u(x,t) - Ru(x,t) - Nu(x,t), \tag{25}$$

Where, $L_x = \dfrac{\partial^2}{\partial x^2}$ and $L_t = \dfrac{\partial}{\partial t}$ are linear operators in the partial differential equation and are easily invertible, $g(x,t)$ is a source term, $Ru(x, t)$ is a remaining linear term and $Nu(x,t)$ is non-linear analytical term expandable in the Adomian polynomials A_n.

After applying the inverse operator L_x^{-1} to both sides of Eq. (25).

$$L_x^{-1} L_x u(x,t) = L_x^{-1} g(x,t) - L_x^{-1} L_t u(x,t) - L_x^{-1} Ru(x,t)$$
$$- L_x^{-1} Nu(x,t), \tag{26}$$

$$u(x,t) = f(x,t) - L_x^{-1} L_t u(x,t) - L_x^{-1} Ru(x,t) - L_x^{-1} Nu(x,t), \tag{27}$$

Here, the function $f(x,t))$ represents the terms arising from $L_x^{-1}(x,t)$ after using the given conditions. $L_x^{-1} = \iint(.)dxdx$ is used as inverse operator for the second order partial differential equation. Similarly, it is used for higher order partial differential equation L_x^{-1} and L_x depend on the order of the partial differential equation.

Adomian Decomposition Method defines the series solution $u(x,t)$ as,

$$u(x,t) = \sum_{n=0}^{\infty} u_n(x,t), \tag{28}$$

$$\sum_{n=0}^{\infty} u_n(x,t) = f(x,t) - L_x^{-1} R \sum_{n=0}^{\infty} u_n(x,t)$$

$$- L_x^{-1} N \sum_{n=0}^{\infty} u_n(x,t), \tag{29}$$

The non-linear term expanding in Adomian polynomials as

$$N \sum_{n=0}^{\infty} u_n(x,t) = \sum_{n=0}^{\infty} A_n, \tag{30}$$

where the components $u_0(x,t)$, $u_1(x,t), u_2(x,t)$, ... are periodically derived as

$$u_0(x,t) + u_1(x,t) + u_2(x,t) + ... = f(x,t)$$

$$- L_x^{-1} R \left(\begin{array}{c} u_0(x,t) + u_1(x,t) \\ + u_2(x,t) + ... \end{array} \right)$$

$$- L_x^{-1} (A_0 + A_1 + ...), \tag{31}$$

To determine the series components $u_0(x,t)$, $u_1(x,t), u_2(x,t)$, ... it is important to note that ADM suggests that the function $f(x,t)$ actually described the zeroth component , $u_0(x,t)$ is usually defined by the function $f(x,t)$ described above.

The formal recursive relation is defined as

$$u_0(x,t) = f(x,t),$$

$$u_1(x,t) = - L_x^{-1} R[u_0(x,t)] - L_x^{-1} [A_0],$$

$$u_2(x,t) = - L_x^{-1} R[u_1(x,t)] - L_x^{-1} [A_1],$$

$$u_3(x,t) = - L_x^{-1} R[u_2(x,t)] - L_x^{-1} [A_2], \text{and so on.} \tag{32}$$

Analysis of Optimal Homotopy Asymptotic Method

For the analysis of OHAM, we consider the boundary value problem as

$$L(u(x,t)) + N(u(x,t)) + G(u(x,t)) = 0, B(u) = 0, \tag{33}$$

Where L a linear operator in the differential equation, N is a non-linear term, x is an independent variable, B is a boundary operator and G is a source term. According to OHAM, we construct a set of equation.

$$[1-p][L\psi(x,t,p) + G(x,t)]$$
$$- H(p)[L\psi(x,t,p) + G\psi(x,t,p) + N\psi(x,t,p)] = 0, \tag{34}$$

$p \in [0,1]$ is an embedding parameter, $H(p) = pc_1 + P^2 c_2 + \ldots\ldots m$, is an auxiliary function and c_1, c_2 are auxiliary constants and $\psi(x,t,p)$ is an unknown function. Obviously, when $p = 0$ and $p = 1$ it holds that:

$$\psi(x,t,p) = u_0(x,t), \psi(x,t,1) = u(x,t), \tag{35}$$

$$\psi(x,t,p,c_i) = u_0(x,t) + \sum_{k \geq 1} u_k(x,t,c_i)p^k, i = 1,2,3\ldots,m, \tag{36}$$

Inserting Eq.(30) in Eq.(28), assembling the similar powers of P and comparing each coefficient of P to zero. The partial differential equations are solved with the given boundary conditions to get $u_0(x,t)$, $u_1(x,t)$, $u_2(x,t)$.

The general solution of Eq.(27) can be written as

$$u^m = u_0(x,t) + \sum_{k=1} u_k(x,t,c_i), \tag{37}$$

The coefficients $c_1, c_2, c_3, \ldots, c_m$ are the functions of x.

Inserting Eq. (31) in Eq.(27), the residual is obtained as:

$$R(x,t,c_i) = L(u^m(x,t,c_i)) + G(x,t) + N(u^m(x,t,c_i)), \tag{38}$$

Numerous methods like Galerkin's Method, Ritz Method, Method of Least Squares and Collocation Method are used to find the optimal values of $c_i, i = 1,2,3,4.....$We apply the Method of Least Squares in our problem as given below:

$$J(c_1,c_2,c_3,....,c_m) = \int_a^b R^2(x,t,c_1,c_2,c_3,...,c_m)dx,$$

(39)

a and b are the constant values taking from domain of the problem.

Auxiliary constants $(c_1, c_2, c_3,....., c_m)$ can be obtained from:

$$\frac{\partial J(c_1,c_2,...,c_m)}{\partial c_1} = \frac{\partial J(c_1,c_2,...,c_m)}{\partial c_2} = ... = \frac{\partial J(c_1,c_2,...,c_m)}{\partial c_m} = 0$$

(40)

Finally, from these auxiliary constants, the approximate solution is well-determined.

The ADM Solution of Lifting Problem

The inverse operator $L_x{}^{-1} = \iint ()dxdx$ is applied on the second order differential Eq. (16) and is according to the standard form of ADM from Eq.(27):

$$u(x,t) = f(x,t) + ML_x^{-1}u + L_x^{-1}\left[\frac{\partial u}{\partial t}\right] - L_x^{-1}\left[\frac{\partial}{\partial t}\left(\frac{\partial^2 u}{\partial x^2}\right)\right],$$

(41)

$$\Theta(x,t) = h(x,t) + P_r L_x^{-1}\left[\frac{\partial \Theta}{\partial t}\right] + B_r L_x^{-1}\left[\begin{array}{c}\left(\frac{\partial u}{\partial x}\right)^2 \\ +\alpha\left(\frac{\partial u}{\partial x}\right)\left(\frac{\partial^2 u}{\partial t \partial x}\right)\end{array}\right],$$

(42)

Summation is used for the series solutions of Eqs. (41,42):

$$\sum_{n=0}^{\infty} u_n = f(x,t) + ML_x^{-1}\left[\sum_{n=0}^{\infty} u_n\right] + L_x^{-1}\left[\frac{\partial}{\partial t}\sum_{n=0}^{\infty} u_n\right]$$

$$- \alpha L_x^{-1}\left[\sum_{n=0}^{\infty} A_n\right],$$

(43)

$$\sum_{n=0}^{\infty} \Theta_n = h(x,t) + P_r L_x^{-1}\left[\frac{\partial}{\partial t}\sum_{n=0}^{\infty} \Theta_n\right] - B_r L_x^{-1}\left[\sum_{n=0}^{\infty} B_n\right]$$

$$- B_r L_x^{-1}\left[\sum_{n=0}^{\infty} C_n\right],$$

(44)

For $n \geq 0$ the Adomian polynomials A_n, B_n and C_n from Eqs.(43,44) are defined as

$$\sum_{n=0}^{\infty} A_n = \frac{\partial}{\partial t}\left(\frac{\partial^2 u}{\partial x^2}\right), \sum_{n=0}^{\infty} B_n = \left(\frac{\partial u}{\partial t}\right)^2,$$

$$\sum_{n=0}^{\infty} C_n = \frac{\partial u}{\partial t}\left(\frac{\partial^2 u}{\partial t \partial x}\right),$$

(45)

In Components form Eqs. (43,44) are derived as:

$$u_0(x,t) + u_1(x,t) + u_2(x,t) + \ldots = f(x,t)$$

$$+ L_x^{-1}\frac{\partial}{\partial t}\left[\begin{pmatrix} u_0(x,t) \\ + u_1(x,t) \\ + u_2(x,t) + \ldots \end{pmatrix}\right]$$

$$+ ML_x^{-1}\begin{pmatrix} u_0(x,t) + u_1(x,t) \\ + u_2(x,t) + \ldots \end{pmatrix}$$

$$- \alpha L_x^{-1}(A_0 + A_1 + A_2 + \ldots),$$

(46)

$$\Theta_0 + \Theta_1 + \Theta_2 + \ldots = h(x,t) + P_r L_x^{-1} \left[\frac{\partial}{\partial t} (\Theta_0 + \Theta_1 + \Theta_2 + \ldots) \right.$$

$$\left. - B_r L_x^{-1} [(B_0 + B_1 + B_2 + \ldots) + \begin{matrix} \alpha(C_0 + C_1 \\ + C_2 + \ldots) \end{matrix} \right], \quad (47)$$

The components of velocity and temperature distribution are obtained by comparing both sides of Eqs. (46,47):

Components of the Lift Problem up to Second Order are:

$$u_0(x,t) = f(x,t) = L_x^{-1} \left(\frac{\partial^2 u_0}{\partial x^2} - S_t \right),$$
$$(48)$$

$$\Theta_0(x,t) = h(x,t) = L_x^{-1} \left(\frac{\partial^2 \Theta_0}{\partial x^2} \right),$$
$$(49)$$

$$u_1(x,t) = L_x^{-1} \left(\frac{\partial u_0}{\partial x} \right) + M L_x^{-1} [u_0] - \alpha L_x^{-1} [A_0],$$
$$(50)$$

$$\Theta_1(x,t) = P_r L_x^{-1} \left(\frac{\partial}{\partial x} (\Theta_0) \right) - B_r L_x^{-1} [B_0 - \alpha(C_0)],$$
$$(51)$$

$$u_2(x,t) = L_x^{-1} \left(\frac{\partial u_1}{\partial x} \right) + M L_x^{-1} [u_1] - \alpha L_x^{-1} [A_1],$$
$$(52)$$

$$\Theta_2(x,t) = P_r L_x^{-1} \left(\frac{\partial}{\partial x} (\Theta_1) \right) - B_r L_x^{-1} [B_1 - \alpha(C_1)],$$
$$(53)$$

Making use of boundary conditions from Eqs.(21,22) in Eqs.(48–53) the zero, first and second components solution are obtained as:

$$u_0(x,t) = 1 + \xi Cos[t\omega] - \left(1 + \xi Cos[t\omega] + \frac{S_t}{2} \right) x + \left(\frac{S_t}{2} \right) x^2,$$
$$(54)$$

$$\Theta_0(x,t) = x,$$
$$(55)$$

$$u_1(x,t) = M\left(\frac{1}{3M}\xi\omega Sin[t\omega] + \frac{S_t}{24} - \frac{1}{3} - \frac{\xi}{3}Cos[t\omega]\right)x$$

$$+ \left(\frac{M}{2} + \frac{1}{2}M\xi Cos[t\omega] - \frac{\xi}{2}\omega Sin[t\omega]\right)x^2$$

$$+ \left(\frac{\xi}{6}\omega Sin[t\omega] - \frac{M}{6} - \frac{1}{6}M\xi Cos[t\omega] - \frac{MS_t}{12}\right)x^3$$

$$+ \left(\frac{MS_t}{24}\right)x^4,$$

(56)

$$\Theta_1(x,t) = B_r \left[\begin{bmatrix} \frac{\xi^2}{3}Cos[t\omega]^2 + \left(\frac{12 + 4S_t + S_t^2}{24}\right) + \\ \left(\frac{S_t + 6}{6}\right)\xi Cos[t\omega] - \frac{\alpha\xi^2\omega}{4}Sin[t\omega] \\ - \left(\frac{S_t + 6}{12}\right)\alpha\omega\xi Sin[t\omega] \end{bmatrix} x \right.$$

$$+ \begin{bmatrix} \frac{\alpha\xi^2\omega}{4}Sin2[t\omega] - \\ \left(\frac{4 + 4S_t + S_t^2}{8}\right) - \left(\frac{S_t + 2}{2}\right)\xi Cos[t\omega] - \\ \frac{\xi^2}{3}Cos[t\omega]^2 + \left(\frac{S_t + 2}{4}\right)\alpha\omega\xi Sin[t\omega] \end{bmatrix} x^2$$

$$+ \left[\frac{2S_t + S_t^2}{6} + \frac{S_t\xi}{3}Cos[t\omega] - \frac{\alpha\xi S_t\omega}{4}Sin[t\omega]\right]x^3$$

$$\left. - \left(\frac{S_t^2}{12}\right)x^4\right],$$

(57)

$$u_2(x,t) = M^2 \left[\frac{1}{45} + \frac{\xi}{45} Cos[t\omega] - \frac{S_t}{240} \right] x$$

$$- \frac{\xi\omega^2}{45} \left[Cos[t\omega] + 15\alpha Cos[t\omega] + \frac{2M}{\omega} Sin[t\omega] \right.$$

$$\left. + \frac{15\alpha M}{\omega} Sin[t\omega] \right] x$$

$$+ \frac{\xi\omega\alpha}{2} [\omega Cos[t\omega] + M Sin[t\omega]] x^2$$

$$+ \frac{M^2}{144} [S_t - 8 - 8\xi Cos[t\omega]] x^3$$

$$+ \frac{\xi\omega^2}{18} \left[Cos[t\omega] - 3\alpha Cos[t\omega] + \frac{2M}{\omega} Sin[t\omega] \right.$$

$$\left. - \frac{3\alpha M}{\omega} Sin[t\omega] \right] x^3$$

$$+ \frac{\xi M^2}{24} \left[\frac{1}{\xi} + Cos[t\omega] - \omega^2 Cos[t\omega] - \frac{2\omega}{M} Sin[t\omega] \right] x^4$$

$$+ \frac{\xi M^2}{240} \left[\begin{array}{c} \frac{4\omega}{M} Sin[t\omega] - \frac{2}{\xi} - 2Cos[t\omega] \\ + 2\omega^2 Cos[t\omega] - \frac{S_t M^2}{\xi} \end{array} \right] x^5$$

$$+ \left[\frac{M^2 S_t}{720} \right] x^6,$$

$$(58)$$

The second term solution for temperature distribution is too bulky, therefore, only graphical representations up to second order are given.

The series solution of velocity distribution up to the second component is as:

$$u(x,t) = u_0(x,t) + u_1(x,t) + u_2(x,t), \qquad (59)$$

Inserting components solutions from Eqs. (54,56,58), in the series solution (59), we have:

$$u(x,t) = 1 + \xi Cos[t\omega]$$

$$
+ \left[\begin{array}{l} \left(\dfrac{M^2\xi}{45} - \dfrac{M\xi}{3} - \dfrac{\omega^2\xi}{45} - \dfrac{\omega\alpha^2\xi}{3} - \xi - \dfrac{\alpha\omega^2\xi}{3} \right) Cos[t\omega] \\[2ex] + \left(\dfrac{\omega\xi}{3} - \dfrac{2M\omega\xi}{45} \right. \end{array} \right.
$$

$$
\left. - \dfrac{M\alpha\omega\xi}{3} \right) Sin[t\omega] - 1 - \dfrac{M}{3} + \dfrac{M^2}{45} - \dfrac{S_t}{2} + \dfrac{MS_t}{24}
$$

$$
\left. - \dfrac{M^2 S_t}{240} \right] x + \left[\left(\dfrac{M\xi}{2} + \dfrac{\omega^2\xi}{2} \right) Cos[t\omega] \right.
$$

$$
+ \left(\dfrac{M\alpha\omega\xi}{2} - \dfrac{\omega\xi}{2} \right) Sin[t\omega] + \dfrac{M}{2} + \left. \dfrac{S_t}{2} \right] x^2
$$

$$
+ \left[\left(\dfrac{\omega^2\xi}{18} - \dfrac{M^2\xi}{18} - \dfrac{M\xi}{6} - \dfrac{\alpha\omega^2\xi}{6} \right) Cos[t\omega] \right.
$$

$$
+ \left(\dfrac{\omega\xi}{6} + \dfrac{M\omega\xi}{9} - \dfrac{M\xi\alpha\omega}{6} \right) Sin[t\omega] + \dfrac{MS_t}{12} - \dfrac{M}{6} + \dfrac{M^2}{18}
$$

$$
\left. - \dfrac{M^2 S_t}{144} \right] x^3 + \left[\left(\dfrac{M^2\xi}{24} - \dfrac{\omega^2\xi}{24} \right) Cos[t\omega] \right.
$$

$$
- \left(\dfrac{M\omega\xi}{12} \right) Sin[t\omega] + \dfrac{MS_t}{24} + \left. \dfrac{M^2}{24} \right] x^4
$$

$$
+ \left[\left(\dfrac{\omega^2\xi}{120} - \dfrac{M^2\xi}{120} \right) Cos[t\omega] + \left(\dfrac{M\omega\xi}{60} \right) Sin[t\omega] - \dfrac{M^2}{120} \right.
$$

$$
\left. - \dfrac{M^2 S_t}{240} \right] x^5 - \left[\dfrac{M^2 S_t}{720} \right] x^6,
$$

$$\tag{60}$$

The OHAM Solution of Lifting Problem

We construct a homotopy for Eqs. (16, 17) from the standard form of OHAM in Eq (34).

According to aforementioned discussion, the zero, first and second components problems are:

$$
p^o : \dfrac{\partial^2 u_0}{\partial x^2} = S_t, \tag{61}
$$

$$
\dfrac{\partial^2 \Theta_0}{\partial x^2} = 0, \tag{62}
$$

$$p^1 : \frac{\partial^2 u_1}{\partial x^2} = -S_t - S_t c_1 - Mc_1 u_0 - c_1 \frac{\partial u_0}{\partial t} + \frac{\partial^2 u_0}{\partial x^2} + c_1 \frac{\partial^2 u_0}{\partial x^2}$$
$$+ \alpha c_1 \frac{\partial}{\partial t}\left(\frac{\partial^2 u_0}{\partial x^2}\right),$$

(63)

$$\frac{\partial^2 \Theta_1}{\partial x^2} = -P_r c_3 \frac{\partial \Theta_0}{\partial t} + B_r c_3 \left(\frac{\partial u_0}{\partial t}\right)^2 + \alpha B_r c_3 \frac{\partial u_0}{\partial x}\left(\frac{\partial^2 u_0}{\partial t \partial x}\right)$$
$$+ \frac{\partial^2 \Theta_0}{\partial x^2} + c_3 \frac{\partial^2 \Theta_0}{\partial x^2},$$

(64)

$$p^2 : \frac{\partial^2 u_2}{\partial x^2} = -S_t c_2 - Mc_2 u_0 - c_2 \frac{\partial u_0}{\partial t} - Mc_1 u_1 - c_1 \frac{\partial u_1}{\partial t} + c_2 \frac{\partial^2 u_0}{\partial x^2}$$
$$+ \alpha c_2 \frac{\partial}{\partial t}\left(\frac{\partial^2 u_0}{\partial x^2}\right) + (1 + c_1)\frac{\partial^2 u_1}{\partial x^2} + \alpha c_1 \frac{\partial}{\partial t}\left(\frac{\partial^2 u_1}{\partial x^2}\right),$$

(65)

$$\frac{\partial^2 \Theta_2}{\partial x^2} = -P_r c_4 \frac{\partial \Theta_0}{\partial t} - P_r c_3 \frac{\partial \Theta_1}{\partial t} + B_r c_4 \left(\frac{\partial u_0}{\partial x}\right)^2$$
$$+ \alpha B_r (c_4 + c_3)\frac{\partial u_0}{\partial x}\left(\frac{\partial^2 u_0}{\partial t \partial x}\right) + 2 B_r c_3 \frac{\partial u_0}{\partial x}\frac{\partial u_1}{\partial x} + c_4 \frac{\partial^2 \Theta_0}{\partial x^2}$$
$$+ (1 + c_3)\frac{\partial^2 \Theta_1}{\partial x^2},$$

(66)

Solving Eqs. (61–66) for zero, first and second components of velocity and temperature profiles by using the corresponding boundary conditions given in Eqs. (21,22) respectively.

$$u_0(x,t) = 1 + \xi Cos[t\omega] - \left(1 + \xi Cos[t\omega] + \frac{S_t}{2}\right)x + \left(\frac{S_t}{2}\right)x^2,$$

(67)

$$\Theta_0(x,t) = x,$$

(68)

$$u_1(x,t,c_1) = c_1 \left[\frac{M}{3} + \frac{M\xi}{3} Cos[t\omega] - \frac{\xi\omega}{3} Sin[t\omega] - \frac{MS_t}{24} \right] x$$

$$- c_1 \left[\frac{M}{2} + \frac{M\xi}{2} Cos[t\omega] - \frac{\xi\omega}{3} Sin[t\omega] \right] x^2$$

$$+ c_1 \left[\frac{M}{6} + \frac{M\xi}{6} Cos[t\omega] - \frac{\xi\omega}{6} Sin[t\omega] \right] x^3$$

$$- c_1 \left[\frac{MS_t}{24} \right] x^4$$

$$(69)$$

$$\Theta_1(x,t) = c_3 \left[-\frac{B_r}{2} - \left(B_r\xi + \frac{S_t B_r \xi}{6} \right) Cos[t\omega] - \frac{B_r \xi^2}{2} Cos[t\omega]^2 \right.$$

$$+ \left(\frac{\alpha\omega B_r \xi}{2} + \frac{\alpha\omega B_r \xi S_t}{12} \right) Sin[t\omega]$$

$$+ \frac{\alpha\omega B_r \xi^2}{2} (Cos[t\omega] Sin[t\omega]) - \frac{B_r S_t}{6} - \frac{B_r S_t^2}{24} \right] x$$

$$+ \left[\frac{B_r}{2} + \left(B_r\xi + \frac{B_r S_t \xi}{6} \right) Cos[t\omega] + \frac{B_r \xi^2}{2} Cos[t\omega]^2 \right.$$

$$- \left(\frac{\alpha\omega B_r \xi}{2} + \frac{\alpha\omega B_r \xi S_t}{4} \right) Sin[t\omega]$$

$$- \frac{\alpha\omega B_r \xi^2}{2} (Cos[t\omega] Sin[t\omega]) + \frac{B_r S_t}{2} - \frac{B_r S_t^2}{8} \right] x^2$$

$$+ \left[\frac{\alpha\omega B_r \xi S_t}{6} Sin[t\omega] - \frac{B_r S_t}{3} - \frac{B_r S_t \xi}{3} Cos[t\omega] \right.$$

$$- \frac{B_r S_t^2}{6} \right] x^3 + \left[\frac{B_r S_t^2}{12} \right] x^4,$$

$$(70)$$

The second term solution for velocity and temperature profiles are too long, therefore, only graphical representations up to second order are given.

The arbitrary constants $c_i, i = 1,2,3,4$ are found by using the residual:

$$R = L(u(x,t,c_i)) + G(u(x,t,c_i)) + N(u(x,t,c_i)), \quad (71)$$

According to Eq.(36), the arbitrary constants for velocity components $u_0(x,t), u_1(x,t), u_2(x,t)$ are

$$c_1 = -0.97616, c_2 = -0.00022.$$

For temperature distribution, the arbitrary constants are

$$c_1 = -0.02275, c_2 = -0.02371, c_3 = -0.93327, c_4 = -0.00447.$$

Formulation of Drainage Problem

The geometry and assumptions of the problem are the same as in the previous problem. Consider, a film of non-Newtonian liquid drains down the vertical belt, the belt is only oscillating and the fluid drain down the belt due to gravity, so the gravity in this case is opposite to the previous case. Therefore, the Stock number is positively mentioned in Eq. (19). The coordinate system is selected same as in previous case. Assuming the flow is unsteady and laminar, fluid shear forces keeps the gravity balanced and the film thickness remains constant.

In drainage problem Eq. (19) reduced as

$$\frac{\partial u}{\partial t} = \frac{\partial^2 u}{\partial x^2} + \alpha \frac{\partial}{\partial t}\left(\frac{\partial^2 u}{\partial x^2}\right) + S_t - Mu, \quad (72)$$

Boundary conditions for drainage problem when belt is only oscillating:

$$u(0,t) = \xi Cos\omega t, \frac{\partial u(\delta,t)}{\partial x} = 0, \quad (73)$$

Using non-dimensional variables from Eq. (14), the boundary conditions (57) of drainage problem are reduced to:

$$u_0(0,t) = \xi Cos\omega t, \frac{\partial u_0(1,t)}{\partial x} = 0, \quad (74)$$

The ADM Solution of Drainage Problem

The model for drainage problem is the same as for the lift problem. The only difference in this problem is that the belt is only oscillating and due to the draining of thin film, stock number is positively mentioned in Eq. (72).

The boundary conditions for temperature distribution are the same as given in Eq. (22) but solution of these components is different. It depends on the different velocity profile of drainage and lift problems. Due to lengthy analytical calculation, solutions of temperature distribution up to first order terms are included whereas the graphical representations up to second order terms are given. Using boundary conditions (22) and (73) into Eqs. (48–53), the component solutions are obtained as:

Components of the Lift Problem up to Second Order are:

$$u_0(x,t) = \xi Cos[t\omega] - \left(\xi Cos[t\omega] + \frac{S_t}{2} \right)x - \left(\frac{S_t}{2} \right)x^2, \tag{75}$$

$$\Theta_0(x,t) = x, \tag{76}$$

$$u_1(x,t) = M \left(\frac{1}{3M} \xi \omega Sin[t\omega] - \frac{S_t}{24} - \frac{\xi}{3} Cos[t\omega] \right)x$$

$$+ \left(\frac{1}{2} M \xi Cos[t\omega] - \frac{\xi}{2} \omega Sin[t\omega] \right)x^2$$

$$+ \left(\frac{\xi}{6} \omega Sin[t\omega] - \frac{1}{6} M \xi Cos[t\omega] + \frac{MS_t}{12} \right)x^3$$

$$- \left(\frac{MS_t}{24} \right)x^4, \tag{77}$$

$$\Theta_1(x,t) = B_r \left[\left[\begin{array}{l} \dfrac{\xi^2}{2} Cos[t\omega]^2 - \dfrac{\alpha\xi^2\omega}{2} Cos[t\omega]Sin[t\omega] \\[2mm] -\dfrac{\xi S_t}{6} Cos[t\omega] + \dfrac{S_t\alpha\xi\omega}{12} Sin[t\omega] + \dfrac{S_t^2}{24} \end{array} \right] x \right.$$

$$\left[\begin{array}{l} -\dfrac{\xi^2}{2} Cos[t\omega]^2 + \dfrac{\alpha\xi^2\omega}{2} Cos[t\omega]Sin[t\omega] \\[2mm] +\dfrac{\xi S_t}{2} Cos[t\omega] - \dfrac{S_t\alpha\xi\omega}{4} Sin[t\omega] - \dfrac{S_t^2}{8} \end{array} \right] x^2$$

$$+ \left[\dfrac{S_t\alpha\xi\omega}{4} Sin[t\omega] \right.$$

$$\left. \left. -\dfrac{\xi S_t}{3} Cos[t\omega] + \dfrac{S_t^2}{6} \right] x^3 - \left[\dfrac{S_t^2}{12} \right] x^4 \right],$$

(78)

$$u_2(x,t) = M^2 \left[\dfrac{\xi}{45} Cos[t\omega] + \dfrac{S_t}{240} \right] x$$

$$-\dfrac{\xi\omega^2}{45} \left[\begin{array}{l} Cos[t\omega] + 15\alpha Cos[t\omega] + \dfrac{2M}{\omega} Sin[t\omega] \\[2mm] +\dfrac{15\alpha M}{\omega} Sin[t\omega] \end{array} \right] x$$

$$+\dfrac{\xi\omega\alpha}{2} [\omega Cos[t\omega] + M Sin[t\omega]] x^2$$

$$-\dfrac{M^2}{144} [S_t + 8\xi Cos[t\omega]] x^3 + \dfrac{\xi\omega^2}{18} [(1-3\alpha) Cos[t\omega]$$

$$+ \left(\dfrac{2M}{\omega} - \dfrac{3\alpha M}{\omega} \right) Sin[t\omega]] x^3$$

$$+\dfrac{\xi M^2}{24} \left[(1-\omega^2) Cos[t\omega] - \dfrac{2}{M} \omega Sin[t\omega] \right] x^4$$

$$+\dfrac{\xi M^2}{240} \left[\dfrac{4\omega}{M} Sin[t\omega] + 2(\omega^2 - 1) Cos[t\omega] \right.$$

$$\left. +\dfrac{S_t M^2}{\xi} \right] x^5 - \left[\dfrac{M^2 S_t}{720} \right] x^6,$$

(79)

The series solution up to the second component is

$$u(x,t) = u_0(x,t) + u_1(x,t) + u_2(x,t),$$

(80)

inserting component solutions from Eqs. (75,77,79), in the series solution (80), we have:

$$u(x,t) = \xi \, Cos[t\omega]$$

$$
+ \left[\left(\frac{M^2\xi}{45} - \frac{M\xi}{3} - \frac{\omega^2\xi}{45} - \xi - \frac{\alpha\omega^2\xi}{3} \right) Cos[t\omega] \right.
$$
$$
\left. + \left(\frac{\omega\xi}{3} - \frac{2M\omega\xi}{45} - \frac{M\alpha\omega\xi}{3} \right) Sin[t\omega] \right.
$$
$$
+ \frac{S_t}{2} - \frac{MS_t}{24} + \frac{M^2 S_t}{240} \right] x + \left[\left(\frac{M\xi}{2} + \frac{\alpha\omega^2\xi}{2} \right) Cos[t\omega] \right.
$$
$$
\left. + \left(\frac{M\alpha\omega\xi}{2} - \frac{\omega\xi}{2} \right) Sin[t\omega] - \frac{S_t}{2} \right] x^2
$$
$$
+ \left[\left(\frac{\omega^2\xi}{18} - \frac{M\xi}{6} - \frac{M^2\xi}{18} - \frac{\partial\xi\omega^2}{6} \right) Cos[t\omega] \right.
$$
$$
\left. + \left(\frac{\omega\xi}{6} + \frac{M\omega\xi}{9} - \frac{M\xi\alpha\omega}{6} \right) Sin[t\omega] + \frac{MS_t}{12} \right.
$$
$$
\left. - \frac{M^2 S_t}{144} \right] x^3 + \left[\left(\frac{M^2\xi}{24} - \frac{\omega^2\xi}{24} \right) Cos[t\omega] \right.
$$
$$
\left. - \left(\frac{M\omega\xi}{12} \right) Sin[t\omega] - \frac{MS_t}{24} \right] x^4
$$
$$
+ \left[\left(\frac{\omega^2\xi}{120} - \frac{M^2\xi}{120} \right) Cos[t\omega] + \frac{M\omega\xi}{60} Sin[t\omega] \right.
$$
$$
\left. + \frac{M^2 S_t}{240} \right] x^5 - \left[\frac{M^2 S_t}{720} \right] x^6,
$$

(81)

The second term solution for temperature distribution are lengthy, therefore, only graphical representations up to second order are given.

The OHAM Solution Of Drainage Problem

From the standard form of OHAM in Eq.(34), we construct a homotopy for Eqs. (72, 20).

According to the aforementioned discussion, the zero, first and second component problems are:

$$p^0 : \frac{\partial^2 u_0}{\partial x^2} = -S_t, \tag{82}$$

$$\frac{\partial^2 \Theta_0}{\partial x^2} = 0, \tag{83}$$

$$p^1 : \frac{\partial^2 u_1}{\partial x^2} = S_t + S_t c_1 - M c_1 u_0 - c_1 \frac{\partial u_0}{\partial t} + (1 + c_1) \frac{\partial^2 u_0}{\partial x^2}$$
$$+ \alpha c_1 \frac{\partial}{\partial t} \left(\frac{\partial^2 u_0}{\partial x^2} \right), \tag{84}$$

$$\frac{\partial^2 \Theta_1}{\partial x^2} = -P_r c_3 \frac{\partial \Theta_0}{\partial t} + B_r c_3 \left(\frac{\partial u_0}{\partial x} \right)^2 + \alpha B_r c_3 \frac{\partial u_0}{\partial x} \left(\frac{\partial^2 u_0}{\partial t \partial x} \right) + \frac{\partial^2 \Theta_0}{\partial x^2}$$
$$+ c_3 \frac{\partial^2 \Theta_0}{\partial x^2}, \tag{85}$$

$$p^2 : \frac{\partial^2 u_2}{\partial x^2} = S_t c_2 - M c_2 u_0 - c_2 \frac{\partial u_0}{\partial t} - M c_1 u_1 - c_1 \frac{\partial u_1}{\partial t} + c_2 \frac{\partial^2 u_0}{\partial x^2}$$
$$+ \alpha c_2 \frac{\partial}{\partial t} \left(\frac{\partial^2 u_0}{\partial x^2} \right) + (1 + c_1) \frac{\partial^2 u_1}{\partial x^2} + \alpha c_1 \frac{\partial}{\partial t} \left(\frac{\partial^2 u_1}{\partial x^2} \right), \tag{86}$$

$$\frac{\partial^2 \Theta_2}{\partial x^2} = -P_r c_4 \frac{\partial \Theta_0}{\partial t} - P_r c_3 \frac{\partial \Theta_1}{\partial t} + B_r c_4 \left(\frac{\partial u_0}{\partial x} \right)^2$$
$$+ \alpha B_r (c_4 + c_3) \frac{\partial u_0}{\partial x} \left(\frac{\partial^2 u_0}{\partial t \partial x} \right) + 2 B_r c_3 \frac{\partial u_0}{\partial x} \left(\frac{\partial u_1}{\partial x} \right)$$
$$+ \alpha B_r c_3 \left(\frac{\partial^2 \Theta_0}{\partial t \partial x} \right) \frac{\partial u_1}{\partial x} + c_4 \frac{\partial^2 \Theta_1}{\partial x^2} + (1 + c_3) \frac{\partial^2 \Theta_1}{\partial x^2}, \tag{87}$$

Solving Eqs. (72,20) by using the corresponding boundary conditions given in Eq. (22) and in Eq. (74). The zero component solution obtained as:

$$u_0 = \xi Cos[t\omega] - \left(\xi Cos[t\omega] + \frac{S_t}{2}\right)x - \left(\frac{S_t}{2}\right)x^2,$$

(88)

$$\Theta_0(x,t) = x,$$

(89)

$$u_1(x,t) = c_1 \left[\frac{M\xi}{3} Cos[t\omega] - \frac{\xi\omega}{3} Sin[t\omega] + \frac{MS_t}{24} \right] x$$

$$+ \left[-\frac{M\xi}{2} Cos[t\omega] + \frac{\xi\omega}{2} Sin[t\omega] \right] x^2$$

$$+ \left[\frac{M\xi}{6} Cos[t\omega] - \frac{\xi\omega}{6} Sin[t\omega] \right] x^3 + \left[\frac{MS_t}{24} \right] x^4,$$

(90)

$$\Theta_1(x,t) = c_3 B_r \left[\left[-\frac{\xi^2}{2} Cos[t\omega]^2 + \frac{a\omega\xi^2}{4} Sin[2t\omega] \right. \right.$$

$$+ \frac{\xi S_t}{6} Cos[t\omega] - \frac{\alpha\omega\xi S_t}{12} Sin[t\omega] - \frac{S_t}{8} \right] x$$

$$+ \left[\frac{\xi^2}{2} Cos[t\omega]^2 - \frac{\alpha\xi^2\omega}{4} Sin[2t\omega] - \frac{S_t\xi}{2} Cos[t\omega] \right.$$

$$+ \frac{S_t\alpha\xi\omega}{4} Sin[t\omega] + \frac{S_t^2}{8} \right] x^2 + \left[\frac{S_t}{3} Cos[t\omega] \right.$$

$$\left. - \frac{S_t\alpha\xi\omega}{6} Sin[t\omega] - \frac{S_t^2}{6} \right] x^3 + \left[\frac{S_t^2}{12} \right] x^4 \right]$$

(91)

The auxiliary constants for the series solution of velocity profile and temperature distribution are respectively:

$$c_1 = -0.98464, c_1 = -0.000017 \, and \, c_1 = -2.4631,$$

$$c_2 = 3.187955, c_3 = -0.780916, c_4 = -0.08042$$

RESULTS AND DISCUSSION

In this article, we have presented and interpreted various results for the thin film flow on a vertical oscillating belt. Figures 1 and 2 show the geometry of lift and drainage velocity profiles. The effect of non-dimensional physical parameter like Stock number S_t, Brinkman number B_r, Prandtl number P_r and Frequency parameter ω in lifting and drainage problems have been discussed in Figs. 3–22. A comparison of the ADM and OHAM solutions for velocity and temperature distribution has been shown in Figs. 3–6 for different values of physical parameters. From these Figs., we conclude that the ADM and OHAM solutions are in quite agreement. The numerical comparison of ADM and OHAM at different time level have been computed in Tables 1–4 for both lift and drainage velocity and temperature profiles respectively. It has been concluded from these tables that absolute error between ADM and OHAM decreases with decrease in time level, while it increases with increase in time level. As the flow of fluid film is subjected to the oscillation as well as translation of the belt, so the velocity and temperature distribution of the fluid film will be high at the surface of the belt comparatively to the residual domain and will decrease gradually for the fluid film away from the surface of the belt. These conclusions have been observed from Tables 1–4 and Figs 7–14. Fig. 15 shows that velocity increases in lift flow when Stock number S_t increases. Physically, it is due to friction which seems smaller near the belt and higher at the surface of the fluid. The velocity of fluid decreases with increasing Stock number in drainage problem shown in Fig 16. Physically, it is due to the fact that increasing Stock number causes the fluids' thickness and reduces its flow. When the flow of fluid is downward in oscillation, velocity increases while it decreases when the flow of fluid is upward. Variations of the magnetic parameter M on lift and drainage velocity profiles have been studied in Figs. 17, 18. Increase in magnetic parameter increases the velocity profile in lift problem but in drainage problem, it is clear that the boundary layer thickness is reciprocal to the transverse magnetic field and velocity decreases as flow progresses towards the surface of the fluid. In lift and drainage velocity profiles, increase in non-dimensional frequency ω changes the direction of fluid flow frequently and steadily converges to a point on the surface of the fluid. If the belt velocity increases with oscillation, the centripetal force decreases and, as a result, velocity of fluid decreases. Figs. 19 and 20 show the effect of Brinkman number B_r for lift and drainage temperature distribution. The temperature distribution increases as the B_r increases and becomes more trampled for higher values of B_r. Figs. 21, 22show the effect of Prandtl number P_r on the lift and drainage temperature distribution. In Eq. (20) Prandtl number P_r is reciprocal to other physical parameters. So increase in Prandtl number P_r decreases the temperature distribution.

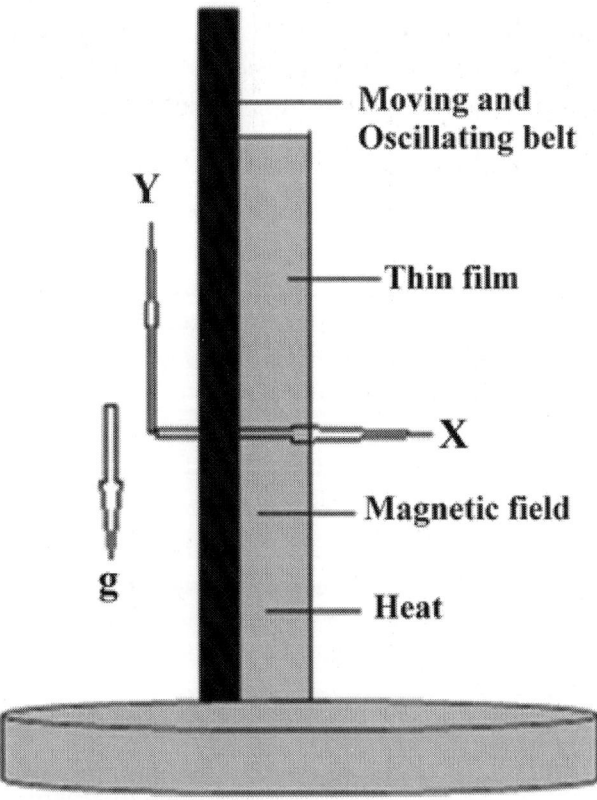

Figure 1: Geometry of the Lift Problem.

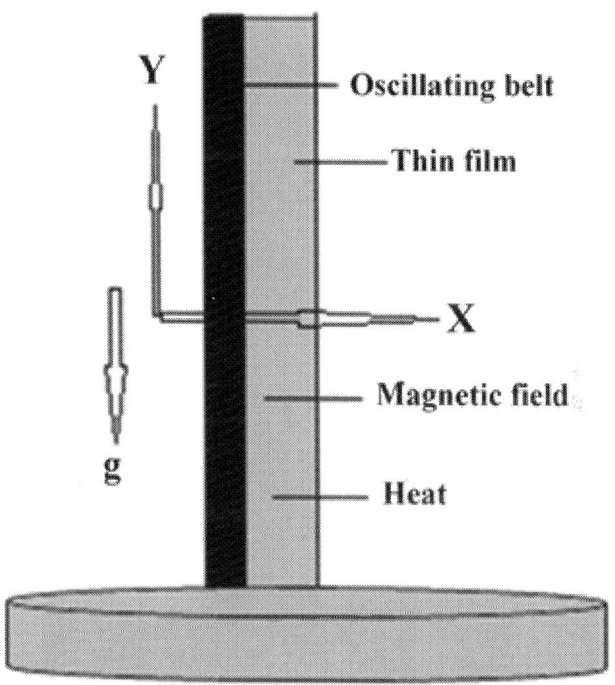

Figure 2: Geometry of the Drainage Problem.

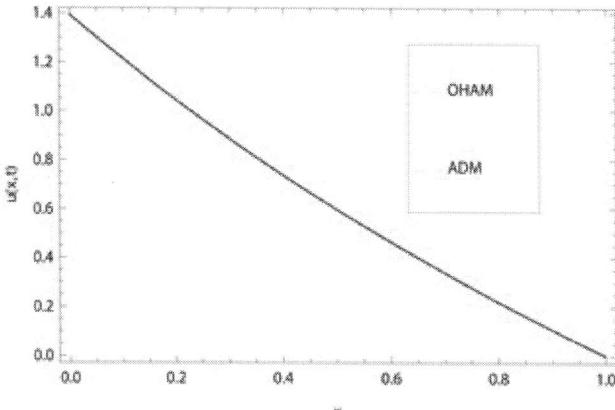

Figure 3: Comparison of Adm And Oham Methods for Lift Velocity

Profile.

$$c_1 = -0.976162, c_2 = -0.00022$$
$$\omega = 0.2, \alpha = 0.02, S_t = 0.5, \ M = 0.5, \xi = 0.4, t = 5, P_r = 0.6, B_r = 4.$$

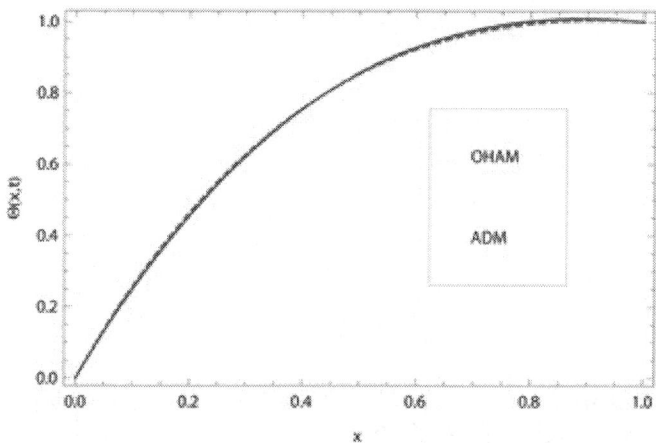

Figure 4: Comparison of Adm and Oham Methods for Lift Temperature Distribution.

$$\omega = 0.2, \alpha = 0.02, S_t = 0.5, M = 0.5, \xi = 0.4, t = 5, \ P_r = 0.6, B_r = 4,$$
$$c_1 = -0.02275, c_2 = -0.023719254, c_3 = -0.933274, c_4 = -0.004472.$$

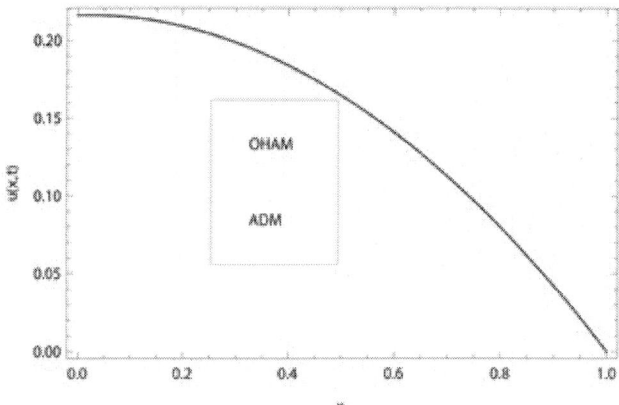

Figure 5: Comparison of Adm And Oham Methods for Drainage Velocity

When

$$c_1 = -0.98464, c_2 = -0.0000174.$$

$$\omega = 0.2, \alpha = 0.02, S_t = 0.5, \ M = 0.5, \xi = 0.4, t = 10, P_r = 0.6, B_r = 4.$$

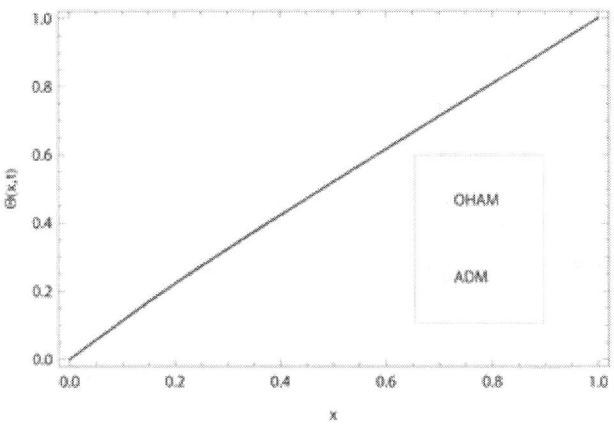

Figure 6: Comparison of Adm and Oham Methods for Temperature Distribution.

$$c_1 = -2.4631, c_2 = -3.187955, c_3 = -0.780916, c_4 = -0.08042.$$

$$\omega = 0.2, \alpha = 0.02, S_t = 0.5, M = 0.5, \xi = 0.4, t = 10, P_r = 0.6, B_r = 4.$$

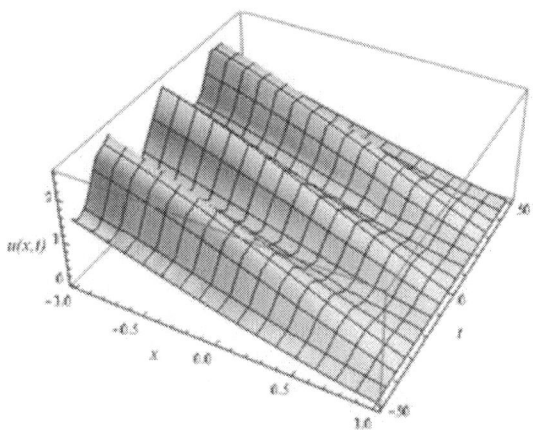

Figure 7: Influence of Different Time Level on Lift Velocity Profile.

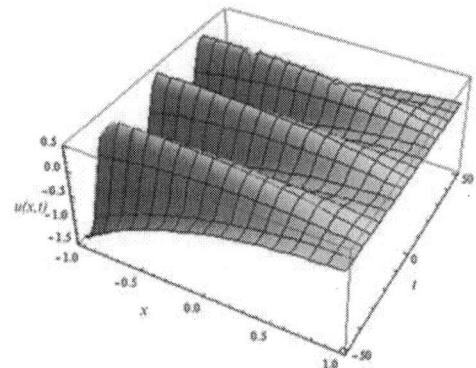

Figure 8: Influence of Different Time Level on Drainage Velocity Profile.

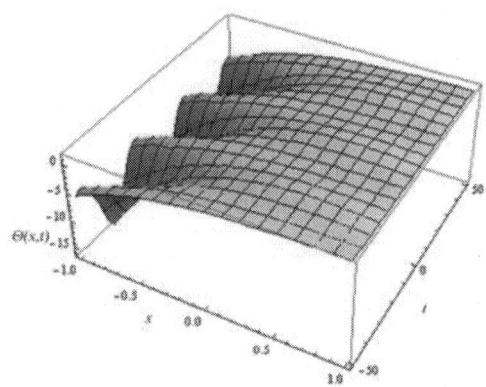

Figure 9: Effect of Different Time Level on Lift Temperature Distribution.

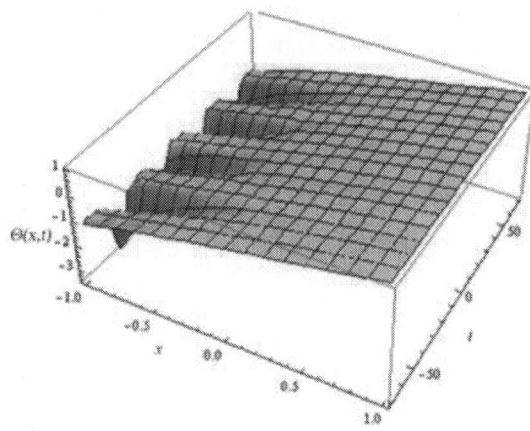

Figure 10: Effect of Different Time Level on Drainage Temperature Distribution.

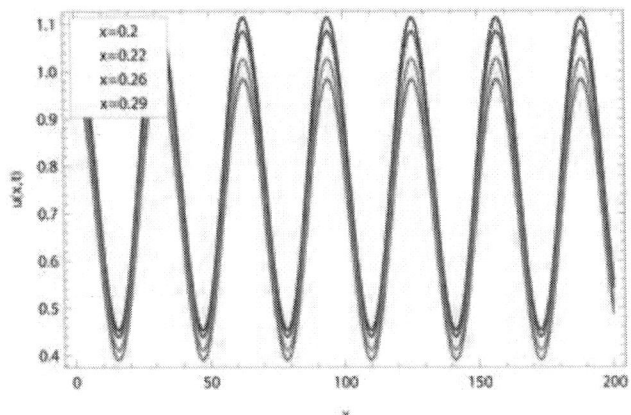

Figure 11: Lift Velocity Distribution at Different Time Level.

$$\omega = 0.2, \alpha = 0.02, S_t = 0.5, M = 0.5, \xi = 0.4.$$

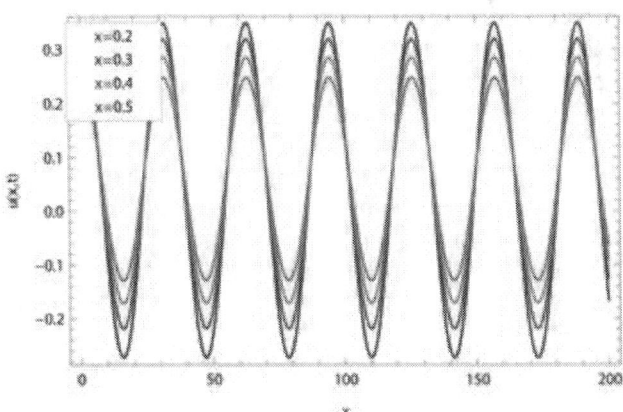

Figure 12: Drainage Velocity Distribution at Different Time Level.

$$\omega = 0.2, \alpha = 0.02, S_t = 0.5, M = 0.5, \xi = 0.4.$$

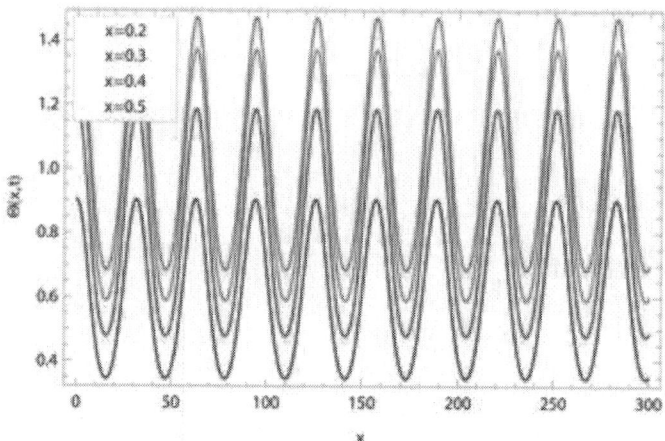

Figure 13: Lift Temperature Distribution of Fluid.

$$\omega = 0.2, \alpha = 0.02, S_t = 0.5, M = 0.5, \xi = 0.4, t = 5, P_r = 0.6, B_r = 4.$$

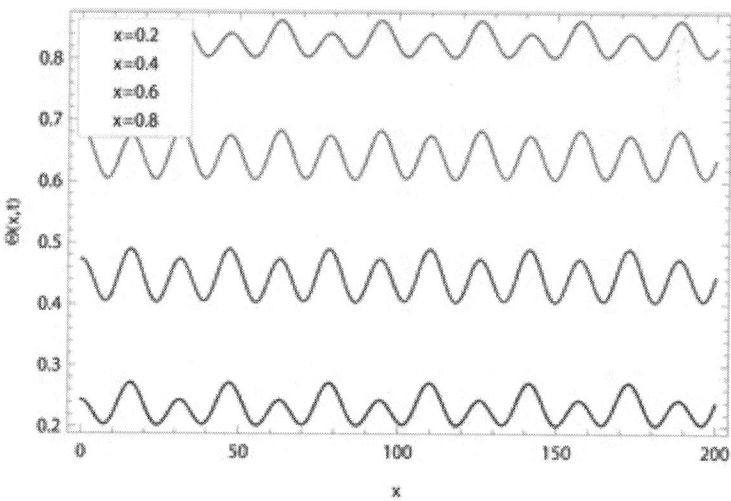

Figure 14: Drainage Temperature Distribution of Fluid.

$$\omega = 0.2, \alpha = 0.02, S_t = 0.5, M = 0.5, \xi = 0.4, t = 5, P_r = 0.6, B_r = 4.$$

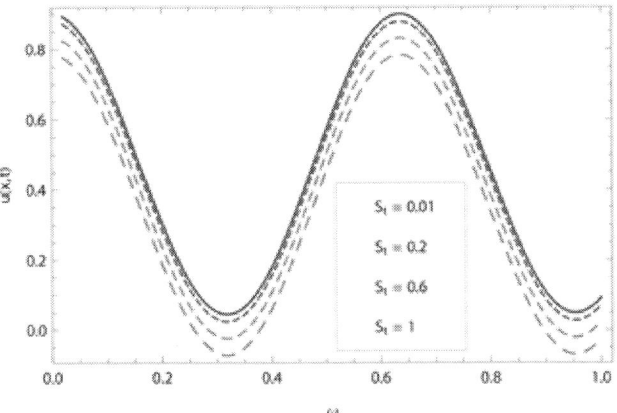

Figure 15: Effect of the Stock Number and Frequency Parameter in Lift Velocity.

$$\alpha = 0.02, M = 0.4, \xi = 0.9, t = 10.$$

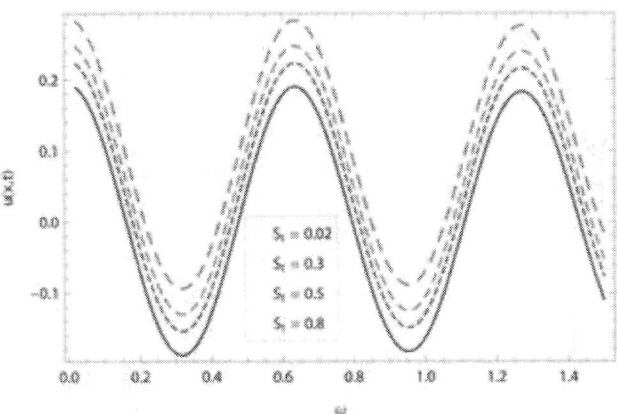

Figure 16: Effect of the Stock Number and Frequency Parameter in Drainage Velocity.

$$M = 0.4, t = 10, \alpha = 0.2, \xi = 0.4.$$

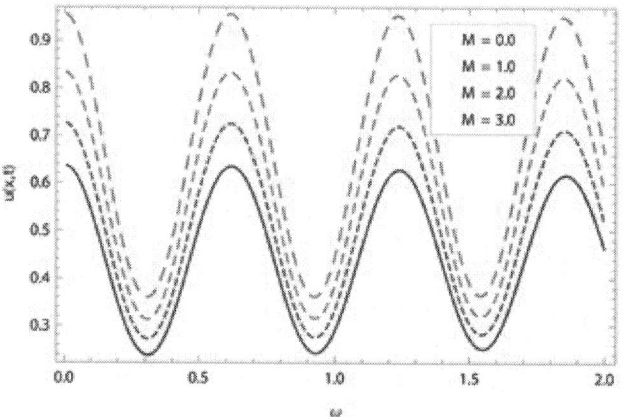

Figure 17: Combined Effect of Magnetic Parameter and Frequency Parameter in Lift Velocity.

$$\alpha = 0.02, S_t = 0.5, \xi = 0.9, t = 10, x = 0.5.$$

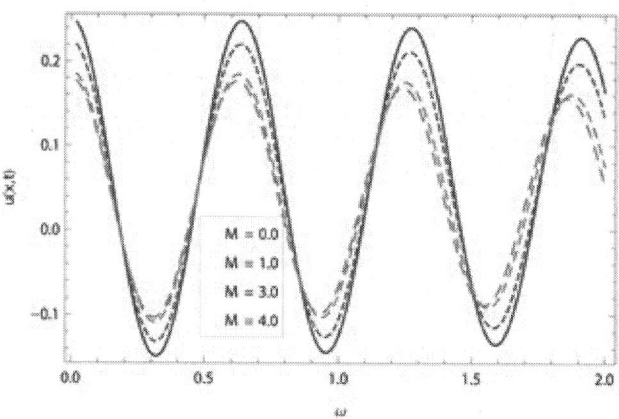

Figure 18: Combined Effect of Magnetic Parameter and Frequency Parameter in Drainage Velocity.

$$\alpha = 0.02, S_t = 0.5, \xi = 0.9, t = 10, x = 0.5.$$

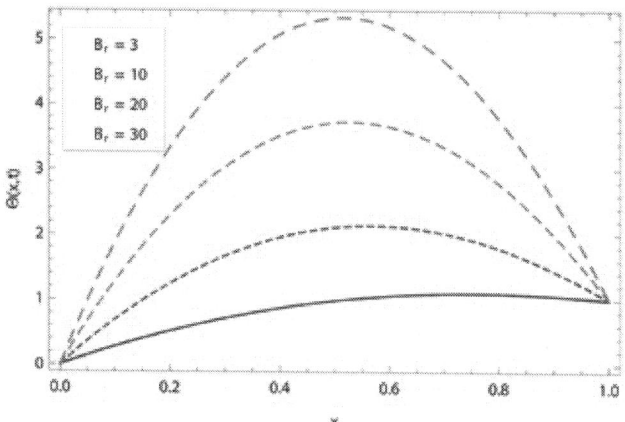

Figure 19: Effect of Brinkman Number in Lift Temperature Distribution.

$$\omega=0.5, \alpha=0.2, S_t=0.5, M=0.5, \xi=0.4, t=10, P_r=0.6.$$

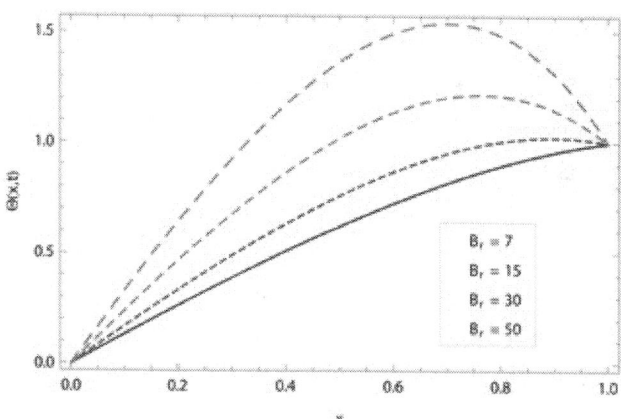

Figure 20: Effect of Brinkman Number in Drainage Temperature Distribution.

$$\omega=0.5, \alpha=0.2, S_t=0.5, M=0.5, \xi=0.4, t=10, P_r=0.6.$$

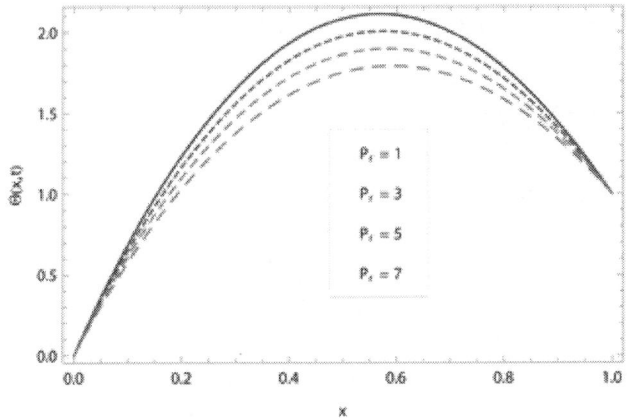

Figure 21: The Effect of Prandtl Number in Lift Temperature Distribution.

$$\omega = 0.5, \alpha = 0.2, S_t = 0.5, M = 0.5, \xi = 0.4, t = 10, B_r = 10.$$

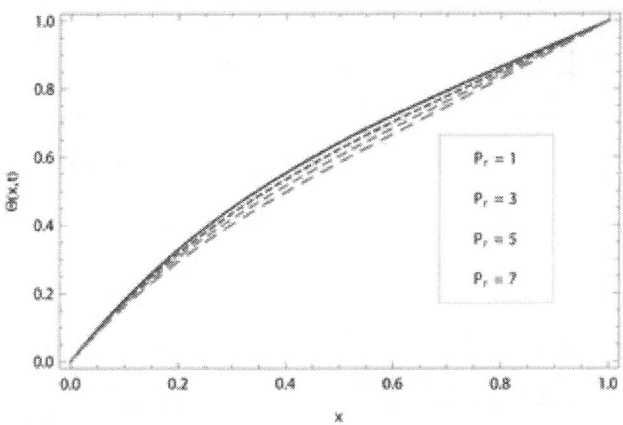

Figure 22: The Effect of Prandtl Number in Drainage Temperature Distribution.

$$\omega = 0.5, \alpha = 0.2, S_t = 0.5, M = 0.5, \xi = 0.4, t = 10, B_t = 10.$$

Table 1: Comparison Of Oham And Adm For Lift Velocity.

x	OHAM	ADM	Absolute Error
0.0	1.392026	1.392026	0
0.1	1.2125619	1.2125847	2.28×10^{-5}
0.2	1.0440308	1.0407481	4.41×10^{-5}
0.3	0.8856109	0.88567161	6.07×10^{-5}
0.4	0.7365279	0.73659884	7.09×10^{-5}
0.5	0.5960522	0.59612614	7.39×10^{-5}
0.6	0.4634961	0.46356567	6.96×10^{-5}
0.7	0.3382102	0.33826889	5.87×10^{-5}
0.8	0.2195811	0.21962343	4.23×10^{-5}
0.9	0.1070279	0.1070279	2.22×10^{-5}
0.10	4.44×10^{-17}	-2.467×10^{-18}	4.68×10^{-17}

Table 2: Comparison of Oham And Adm for Lift Temperature Distribution.

x	OHAM	ADM	Absolute Error
0.0	0	0	0
0.1	0.2491118	0.2561266	7.01×10^{-3}
0.2	0.4552903	0.4634234	8.13×10^{-3}
0.3	0.6225271	0.6282834	5.75×10^{-3}
0.4	0.7546329	0.7563998	1.76×10^{-3}
0.5	0.8552365	0.8528328	2.40×10^{-3}
0.6	0.9277853	0.9220711	5.71×10^{-3}
0.7	0.9755444	0.9680901	7.45×10^{-3}
0.8	1.0015971	0.9944065	7.19×10^{-3}
0.9	1.0088431	1.0041281	4.71×10^{-3}
0.10	1.0000000000000004	0.99999999999	1.16×10^{-15}

Table 3: Comparison of Oham And Adm for Drainage Velocity Profile.

x	OHAM	ADM	Absolute Error
0.0	0.2162092	0.2162092	0
0.1	0.21480638	0.21480528	1.11×10^{-6}
0.2	0.20897722	0.20897538	1.84×10^{-6}
0.3	0.19867913	0.19867691	2.22×10^{-6}
0.4	0.18393209	0.18392977	2.32×10^{-6}
0.5	0.16473082	0.16472863	2.18×10^{-6}
0.6	0.14104521	0.14104332	1.89×10^{-6}
0.7	0.11282061	0.16472863	1.48×10^{-6}
0.8	0.07997793	0.07997691	1.01×10^{-6}
0.9	0.04241369	0.04241319	5.16×10^{-7}
1.0	3.778×10^{-18}	3.084×10^{-19}	3.46×10^{-18}

CONCLUSION

In this article, we have modeled the thin film flow of unsteady second grade fluid on a vertical oscillating belt. The belt is oscillating and translating for lift velocity distribution while belt is only oscillating for drainage velocity distribution in the form of partial differential equation. Both problems have been solved analytically by ADM and OHAM. The comparison of ADM and OHAM has been derived graphically and numerically. We have concluded that the velocity and temperature distribution of the fluid film will be high at the surface of the belt comparatively to the residual domain and will decrease gradually for the fluid film away from the surface of the belt. Expression for velocity and temperature fields have been resulted and sketched. The effects of physical parameters have been sketched and discussed.

REFERENCES

1. Adomian G (1992) A Review of the Decomposition Method and Some Recent Results for Non-Linear Equations, Math Comput. Model. 13: 287–299.

2. Adomian G (1994) Solving Frontier Problems of Physics: the Decomposition Method, Kluwer Academic Publishers

3. Ali F, Khan I, Shafie S (2014) Closed Form Solutions for Unsteady Free Convection Flow of a Second Grade Fluid over an Oscillating Vertical Plate. PLoS ONE 9(2): e85099 doi:10.1371/journal.pone.0085099.

4. Ancey C (2007) Plasticity and Geophysical Flows: a Review. J. Non-Newt.Fluid Mech (142): 4–35. doi: 10.1016/j.jnnfm.2006.05.005

5. Cherruault Y (1990) Convergence of Adomian'sMethod. Kybernotes18(20): 31–38.

6. Cherruault Y, Adomian G (1993) Decomposition Methods: a New Proof of Convergence.Math. Comput. Modelling 8(12): 103–106. doi: 10.1016/0895-7177(93)90233-o

7. Fetecau C, Fetecau C (2005) Starting Solutions for Some Unsteady Unidirectional Flows of a Second Grade Fluid. Int. J. Eng. Sci. (43): 781–789. doi: 10.1016/j.ijengsci.2004.12.009

8. Fetecau C, Fetecau C (2006) Starting Solutions for the Motion of a Second Grade Fluid Due to Longitudinal and Torsional Oscillations of a Circular Cylinder, Int.J. Eng. Sci. 44: 788–796. doi: 10.1016/j.ijengsci.2006.04.010

9. Griffiths RW (2000) The Dynamics of Lava Flows. Annu. Rev. Fluid Mech. 32: 477–518. doi: 10.1146/annurev.fluid.32.1.477

10. Gul T, Islam S, Shah RA, Khan I, Shafie S (2014) Thin Film Flow in MHD Third Grade Fluid on a Vertical Belt with Temperature Dependent Viscosity. PLoS ONE 9(6): e97552 doi:10.1371/journal.pone.0097552.

11. Gul T, Shah RA, Islam S, Arif M (2013) MHD Thin Film Flows of a Third Grade Fluid on a Vertical Belt With Slip Boundary Conditions. J Appl Math: Article ID 707286 pp 14.

12. Han CD, Rao D (1978) The Rheology of Wire Coating Extrusion. Polymer engineering and science 18(13): 1019–1029. doi: 10.1002/pen.760181309

13. Kamran AM, RahimMT, Haroon T, Islam S, Siddiqui AM (2012) Thin-Film Flow of Johnson-SegalmanFluids for Lifting and Drainage Problems,. Appl. Math.Comput (218): 10413–10428. doi: 10.1016/j.amc.2012.03.095

14. Khan NA, Aziz S, Khan NA (2014) Numerical Simulation for the Unsteady MHD Flow and Heat Transfer of Couple Stress Fluid over a Rotating Disk. PLoS ONE 9(5): e95423 doi:10.1371/journal.pone.0095423.

15. Liao SJ (2003) Beyond Perturbation: Introduction to Homotopy Analysis Method. Chapman & Hall/CRC Press, Boca Raton.

16. Mabood F, Khan WA, Ismail AIM (2013) Optimal Homotopy Asymptotic Method for Flow and Heat Transfer of a Viscoelastic Fluid in an Axisymmetric Channel with a Porous Wall. PLoS ONE 8(12): e83581 doi:10.1371/journal.pone.0083581.

17. Marinca V, Herisanu NN (2008) Application of Optimal Homotopy Asymptotic Method for Solving Non-Linear Equations Arising in Heat Transfer,. Int. Communications in heat and mass.Trans 35: 710–715. doi: 10.1016/j.icheatmasstransfer.2008.02.010

18. Marinca V, Herisanu NN, Nemes I (2008) Optimal Homotopy Asymptotic Method with Application to Thin Film Flow. Cent. Eur. J. Phys 6(3): 648–653. doi: 10.2478/s11534-008-0061-x

19. Marinca V, HerisanuNN, Constantin B, Bogdan M (2009) An Optimal Homotopy Asymptotic Method Applied to the Steady Flow of a Fourth Grad Fluid Past a Porous Plate. Appl. Math. Lett (22): 245–251. doi: 10.1016/j.aml.2008.03.019

20. Miladinova S, Lebon G, Toshev E (2004) Thin Film Flow of a Power Law Liquid Falling Down an Inclined Plate. J.Non-Newtonian Fluid Mech(122): 69–70. doi: 10.1016/j.jnnfm.2004.01.021

21. Moli Z, Shaowei W and Shoushui W,Transient Electro-Osmotic Flow of Oldroyd-B Fluids in a Straight Pipe of Circular Cross Section, Journal of Non-Newtonian Fluid Mechanics 201 (2013) 135–139.

22. Moli Z, Shaowei W, Qiangyong Z (2014) Onset of Triply Diffusive Convection in aMaxwell Fluid Saturated Porous Layer, Applied Mathematical Modelling. 38: 2345–2352. doi: 10.1016/j.apm.2013.10.053

23. Qasim M (2013) Heat and Mass Transfer in aJeffrey Fluid over a Stretching Sheet With Heat Source. Alex Eng J 52: 571–575. doi: 10.1016/j.aej.2013.08.004

24. Qasim M, Hayat T, Obaidat S (2012) Radiation Effect on the Mixed Convection Flow of a Viscoelastic Fluid Along an Inclined Stretching Sheet. Z Naturforsch 67: 195–202. doi: 10.5560/zna.2012-0006

25. Qasim M, Khan ZH, Khan WA, Ali SI (2014) MHD Boundary Layer Slip Flow and Heat Transfer of Ferrofluid along a Stretching Cylinder with Prescribed Heat Flux. PLoS ONE 9(1): e83930 doi:10.1371/journal.pone.0083930.

26. S. Wangand W. Tan,Stability Analysis of Double-Diffusive Convection of Maxwell Fluid in a Porous Medium Heated From Below, Phys. Lett. A 372 (2008) 3046–3050.

27. Samiulhaq, Ahmad S, Vieru D, Khan I, Shafie S (2014) Unsteady Magnetohydrodynamic Free Convection Flow of a Second Grade Fluid in a Porous Medium with Ramped Wall Temperature. PLoS ONE 9(5): e88766 doi:10.1371/journal.pone.0088766.

28. Shah AR, Islam S, Siddiqui AM, Haroon T (2011) OHAM solution of unsteady second grade fluid in wire coating analysis. J. KSIAM (15): 201-222.

29. Squires TM, Quake SR (2005) Micro-fluidics: Fluid Physics at the Nanoliter Scale. Rev.Mod.Phys (77): 977-1024. doi: 10.1103/revmodphys.77.977

30. Stone HA, Strook AD, Ajdari A (2004) Engineering Flows in Small Devices. Annu. Rev.Fluid Mech (36): 381-411. doi: 10.1146/annurev.fluid.36.050802.122124

31. Wang S, Tan W,Stability Analysis of Soret-Driven Double-Diffusive Convection of Maxwell Fluid in a Porous Medium. Int. J. Heat Fluid Flow 32 (2011)88-94

32. Wazwaz A, Adomian M (2005) Decomposition Method for a Reliable Treatment of the Bratu-Type Equations, Appl. Math. Comput. (166): 652-663. doi: 10.1016/j.amc.2004.06.059

33. Wazwaz AM (2005) Adomian Decomposition Method for a Reliable Treatment of the Emden–Fowler Equation, Appl. Math. Comput. (161): 543-560. doi: 10.1016/j.amc.2003.12.048

34. Weinstein SJ, Ruschak KJ (2004) Coating Flows. Annu. Rev.Fluid Mech (36): 29-53. doi: 10.1146/annurev.fluid.36.050802.122049

35. Zhao M, Zhang Q and Wang S (2014) Linear and Nonlinear Stability Analysis of Double Diffusive Convection in a Maxwell Fluid Saturated Porous Layer with Internal Heat Source. J ApplMaths, Article ID 489279.

CITATION

Gul T, Islam S, Shah RA, Khan I, Khalid A, et al. (2014) Heat Transfer Analysis of MHD Thin Film Flow of an Unsteady Second Grade Fluid Past a Vertical Oscillating Belt. PLoS ONE 9(11): e103843. doi:10.1371/journal.pone.0103843

CHAPTER 6

Condensation Heat Transfer in Horizontal Non-Circular Microchannels

Hicham El Mghari[1,2], Mohamed Asbik[1], Hasna Louahlia-Gualous[2]

[1]LP2MS, URAC08, Université Moulay Ismaïl, Faculté des Sciences, Meknès, Maroc
[2]LUSAC, Université de Caen Basse Normandie, Saint Lô, France

Keywords

Condensation; Microchannel; Numerical simulation; Capillary regime; Heat transfer

ABSTRACT

This investigation contributes to a better understanding of condensation heat transfer in horizontal non-circular microchannels. For this purpose, the conservation equations of mass, momentum and energy have been numerically solved in both phases (liquid and vapor), and all the more, so the film thickness analytical expression has been established. Numerical results relative to variations of the meniscus curvature radius, the condensate film thickness, the condensation length and heat transfer coefficients, are analyzed in terms of the influencing physical and geometrical quantities. The effect of the microchannel shapes on the average Nusselt number is highlighted by studying condensation of steam insquare, rectangular and equilateral triangular microchannels with the same hydraulic diameter of 250 μm.

INTRODUCTION

Understanding the heat transfer behavior of condensation flow in microchannels is important for a broad variety of engineering applications. Although there have been a number of investigations on boiling flow in microchannels, there are relatively little experimental data and theoretical analyses relative to condensation processes available in the literature, especially, for condensation inside a noncircular microchannels. In the chapter 6 of the reference [1], an overview of minichannels and microchannels condensation has been exposed.

Most of the physical and mathematical models that focused on annular condensation heat transfer in circular channel were developed in the previous works. Begg et al. [2] studied annular film condensation in a small circular tube to predict the shape of the liquid-vapor interface along a miniature tube leading to the complete condensation phenomena in small diameter tubes. LouahliaGualous and Asbik [3] conducted a numerical model predicting heat transfer for condensation of pure refrigerant and binary mixture in a mini-tube. Miscevic et al. [4] developed a stationary condensation capillary flow model based on the separate flow approach by taking into account the coupling between a cylindrical interface and a hemispherical interface. Recently, Ribeiro et al. [5] experimentally investigated the thermal-hydraulic performance of microchannel condensers using three different copper metal foams structures with distinct pore densities and porosities (0.893 and 0.947) as enhanced surfaces on the air-side. Their results are compared with the conventional condenser surface. El Achkar et al. [6] investigated the experimental heat transfer in the isolated bubbles zone of a transparent circular cross-section micro condenser. The evolution of vapor quality was experimentally determined by using the image processing. The energy balance was then used to calculate the temperature of the liquid in the isolated bubbles zone, showing that liquid and vapor were not in thermal equilibrium. The sensible heat transfers and latent heat transfers were then compared. However, a fundamental understanding of local mechanisms of heat and mass transfer cannot be accomplished using multichannels because the mass flow rates for each microchannel and condensate flow regime are unknown. Also, condensation heat transfer and pressure drop depend on the corresponding structure of the two phase flow (mist flow, annular flow, bubbly flow, or slug/plug flow) as shown by Odaymet et al. [7]. Three main condensation flows were identified in a small circular tube: the annular flow, the intermittent or elongated bubbles flow, and the spherical bubbles flow by Louahlia-Gualous and Mecheri [8]. Annular flow is especially found to be one of the dominant condensation flows in microchannelsas shown by Odaymet and LouahliaGualous [9] and Quan et al. [10].

On the other hand, various theoretical models have been proposed to predict the local heat transfer related to the condensation annular flow in non-circular channel where the surface tension plays a predominant effect on the condensate flow, more specifically, in the channel corners. Indeed, Zhao and Liao [11] analyzed annular film condensation heat transfer inside vertical mini triangular channel using three zones: the thin liquid film flow on the sidewall, the condensate flow in the corners, and the vapor core flow in the center. Wu and Cheng [12] carried out a simultaneous visualization and measurement experiment to perform condensation flow patterns of steam flowing through an array of trapezoidal silicon microchannels with a hydraulic diameter of 82.8 μm. Wang et al. [13] and Wang & Rose [14] proposed a theoretical model for condensation annular flow in a horizontal square and equilateral triangular channel with hydraulic diameter ranging from 0.5 to 5 mm by taking into account the effects of gravity, surface tension, and interfacial shear stress. They obtained the local heat transfer coefficient for refrigerants R134a, R22, R152a, CO2, propane, ammonia, and R410a by assuming that the channel wall temperature is uniform. Furthermore, they proposed one correlation for condensation heat transfer in the square and triangular microchannels in which surface tension and viscosity are the predominant parameters controlling condensate film thickness. Additionally, in references ([15-18]) for which the contents are not detailed here, readers could find more information about the use of various non-circular microchannel shapes.

The main purpose of the present work is to determine the heat transfer coefficient during the steam condensation inside horizontal various non-circular microchannels (rectangle, square, or equilateral triangle). Indeed, the classical mathematical model of the annular condensation flow is retained in both phases (liquid and vapor). An appropriate numerical method is used to solve the differential equations system obtained from different conservation equations (mass, momentum and energy). To compute the heat transfer coefficient, a new and simple geometrical method is used to express the condensate film thickness.

PHYSICAL MODEL

The physical model investigated in this paper is illustrated on the **Figure 1**. It concerns, a horizontal non circular microchannel with a hydraulic diameter D, which is cooled with a uniform wall heat flux density q. At the microchannel inlet, the

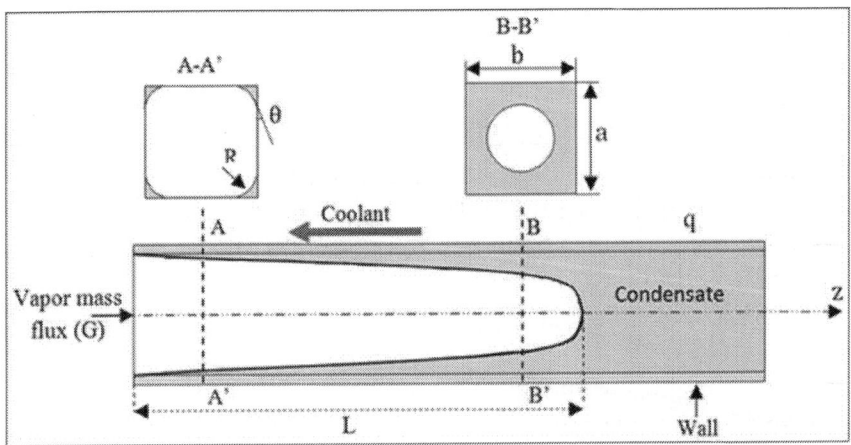

Figure 1: Schematic representation of the annular condensation in a rectangular microchannel.

vapor (steam) mass flux, the pressure and temperature of the steam is given at the saturated state. The vapor condenses inside the microchannel having wall temperature lower than vapor saturation temperature. The film thickness on the heat exchange surface varies along the axial direction with the vapor quality.

When total condensation occurs, the end of the condensation zone has a hemispherical meniscus. Its length L is one of the unknown parameters in the physical model. The condensate film flows along the axial direction under effects of the pressure, surface tension, and shear stress.

MATHEMATICAL FORMULATION

Hypotheses

The mathematical formulation of the problem is based on the following principal assumptions:

- For liquid and vapor phases, thermophysical properties are assumed to be constant.
- In the microchannel, the flow is supposed to be steady-state, laminar, one-dimensional and axis-symmetrical.
- The free surface of the condensate film is smooth.
- Gravity forces are negligible compared to the effects of surface tension.
- The heat transfer from the cooling fluid to the condensate flow is assumed to be one-dimensional.
- In the condensate, the temperature profile is supposed linear.
- The saturation temperature of the vapor is assumed to remain constant along the microchannel.

Conservation Equations

The modeling approach developed here describes the liquid and vapor phases separately. The governing equations are used in the Cartesian coordinates as shown in the Figure 1.

Mass Conservation

The average parameters over a cross-section are used in liquid and vapor phases of the condensation flow respecting continuity conditions at the liquid-vapor interface. The equation of the mass conservation can be written for each local cross section as follows:

$$\frac{1}{A}\frac{\partial}{\partial z}\left(\rho_l U_l\left(1-\alpha\right)A\right)=-\Gamma$$

(1)

$$\frac{1}{A}\frac{\partial}{\partial z}\left(\rho_v U_v \alpha A\right)=\Gamma$$

(2)

where Γ represents the volumic rate of phase change. By convention, its sign is negative for condensation and positive for evaporation. So, we can write:

$$\Gamma=-d\dot{m}_L/dz=d\dot{m}_v/dz$$

(3)

A_ξ and U_ξ are respectively the microchannel cross section, the density and the axial velocity. The subscript ξ refers to the considered phase (ξ = v or L).

Momentum Conservation

For a fixed position z along the microchannel, the condensate film thickness in noncircular microchannels is much thicker into the microchannel corners than elsewhere, especially; at the internal considered circumference because of the surface tension effect. This is the reason that the axial flow in the film region between the corners is neglected.

From the forces balance illustrated on the Figure 2, the momentum conservation equations in the control volume of the length dz, are given by the equations (4) and (5). These relationships are essentially expressed in terms ofthe interfacial shear at the liquid-vapor interface, the shear wall friction at the liquid-wall contact surface, and the pressure forces on the liquid area. In both phases, all the physical properties are assumed to be constant, and the influences of the gravity and the buoyancy forces are neglected. So, in the liquid and vapor phases, the momentum conservation equations are:

• In the liquid phase:

$$-A_L \frac{\mathrm{d}P_L}{\mathrm{d}z} + \tau_{vL} \frac{\mathrm{d}S_{vL}}{\mathrm{d}z} - \tau_w \frac{\mathrm{d}S_{Lw}}{\mathrm{d}z} - U_L \frac{\mathrm{d}\dot{m}_L}{\mathrm{d}z} = \dot{m}_L \frac{\mathrm{d}U_L}{\mathrm{d}z}$$

(4)

• In the vapor phase:

$$-A_v \frac{\mathrm{d}P_v}{\mathrm{d}z} + \tau_{vL} \frac{\mathrm{d}S_{vL}}{\mathrm{d}z} - U_v \frac{\mathrm{d}\dot{m}_v}{\mathrm{d}z} = \dot{m}_v \frac{\mathrm{d}U_v}{\mathrm{d}z}$$

(5)

A_ξ, A_ξ and U_ξ are respectively the microchannel cross-section, the pressure and the axial velocities in the phase $\xi(v\,or\,l)$. Furthermore, we indicate that $\mathrm{d}S_{vL}$ is the liquid–vapor interface surface along dz, $\mathrm{d}S_{Lw}$ is the wet heat exchange surface along dz, and τ_{vL} is the shear stress at liquid–vapor interface, τ_w is the shear stress at the microchannel heat exchange surface.

Energy Conservation
The local energy equation in the liquid phase can written as:

$$q\mathcal{P}z = \dot{m}_L(z)h_{fg}$$

(6)

The total energy equation defined in the length of the total condensation zone as:

$$q\mathcal{P}L = \dot{m}_{v,in}h_{fg}$$

(7)

where q is the heat flux density, $m_L(z)$ is the local liquid mass flow rate ρ is the microchannel perimeter, and z is the microchannel abscissa.

Curvature Radius Expression

To express the curvature radius derivative, we need to use the Laplace–Young equation:

$$\frac{d P_L}{d z} = \frac{d P_v}{d z} + \frac{\sigma}{R^2}\frac{d R}{d z} \tag{8}$$

Combining equations (1)-(8), we get:

$$\frac{d R}{d z} = \frac{0.5 f_{vL}\rho_v P_{vL} U_v^2 \left[1 + \dfrac{A_L}{A_v}\right] - 0.5 f_L \rho_L P_{Lw} U_L^2 - \dfrac{2qPA_L}{h_{fg}}\left[\dfrac{U_v}{A_v} + \dfrac{U_L}{A_L}\right]}{\sigma\dfrac{A_L}{R^2} + 2\dfrac{\rho_v U_v^2 A_L^2}{A_v R} - \dfrac{2\rho_L A_L}{R} U_L^2} \tag{9}$$

Dimensionless Equations

To establish the dimensionless equations, the following variables are used:

$$R^* = \frac{R}{D_h}, z^* = \frac{z}{D_h}, L^* = \frac{L}{D_h}, \bar\rho = \frac{\rho_v}{\rho_L}, P^* = \frac{P}{D_h}, P^* = \frac{PD_h}{2\sigma}, U_v^* = \frac{A\rho_v U_v}{\dot m_{v,in}}, U_L^* = \frac{A\rho_L U_L}{\dot m_{v,in}} \tag{10}$$

Then the dimensionless form of the derivative curvature radius, the velocity gradients and the pressure gradients in both phases, are given by:

$$\frac{dR^*}{dz^*} = \frac{0.5 f_{vL} P_{vL}^* U_v^{*2}\left[1 + \dfrac{A_L^*}{A_v^*}\right] - 0.5 f_L \rho^* P_{Lw}^* U_L^{*2} - 2Bo P^* A_L^*\left[\rho^*\dfrac{U_L^*}{A_L^*} + \dfrac{U_v^*}{A_v^*}\right]}{\dfrac{A_L^* \rho^*}{Re\,Ca} + \dfrac{2 A_L^{*2} U_v^{*2}}{A_v^* R^*} - \dfrac{2\rho^* A_L^* U_L^{*2}}{R^*}} \tag{11}$$

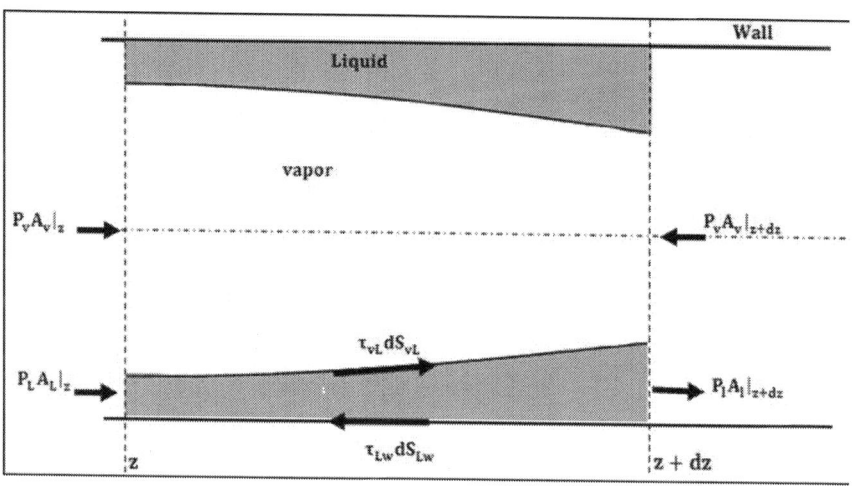

Figure 2: Forces balance for two-phase flow in a horizontal microchannel.

$$\frac{dU_L^*}{dz^*} = \frac{Bo\,P^*}{A_L^*} - 2\frac{U_L^*}{R^*}\frac{dR^*}{dz^*}$$

(12)

$$\frac{dU_v^*}{dz^*} = -\frac{Bo\,P^*}{A_v^*} - 2\frac{A_L^*}{A_v^*}\frac{U_v^*}{R^*}\frac{dR^*}{dz^*}$$

(13)

$$\frac{dP_L^*}{dz^*} = \frac{1}{4}\frac{f_{vL}Ca\,Re\,P_{vL}^*U_v^{*2}}{\bar{\rho}A_L^*} - \frac{1}{4}\frac{f_L Ca\,Re\,P_{Lw}^*U_L^{*2}}{A_L^*}$$

$$-\frac{Bo\,Ca\,Re\,P^*U_L^*}{A_L^*} + \frac{Ca\,Re\,U_L^{*2}}{R^*}\frac{dR^*}{dz^*}$$

(14)

$$\frac{dP_v^*}{dz^*} = -Ca\,Re\,U_v^*\frac{dU_v^*}{dz^*} + \frac{Ca\,Re}{A_v^*}\frac{dA_v^*}{dz^*}$$

$$+\frac{Ca\,Re\,U_v^{*2}}{R^*}\frac{A_L^*}{A_v^*}\frac{dR^*}{dz^*} - \frac{1}{4}\frac{f_{vL}Ca\,Re\,P_{vL}^*U_v^{*2}}{\bar{\rho}A_v^*}$$

(15)

$$Bo = q/Gh_{fg}\,, \quad Ca = \frac{\mu_l G}{\sigma \rho_L}\,,$$

$$Re = \frac{\dot{m}_{v,in}D_h}{A\mu_L}\,, \quad \dot{m}_{v,in} = GA$$

are respectively the boiling number, the capillary number, the vapor Reynolds number and the vapor mass flux.

Definitions of other parameters appearing in the above relations (11-15) are such that:

1) The liquid friction factor for laminar is given by:

$$f_L = CRe_L^{-1}$$
(16)

C is the Poiseuille number given in [19].

For turbulent flow, the liquid friction coefficient is determined from the Blasius equation [1]:

$$f_L = 0.0791 Re_L^{-0.25}$$
(17)

In this expression, the liquid Reynolds number Re_L is calculated assuming the liquid single phase:

$$Re_L = \frac{\rho_L U_L D_L}{\mu_L}, \quad D_L = \frac{4A_L}{P_{Lw}}$$
(18)

2) The interfacial frictional coefficient taking into account the effect of the condensation process on the interfacial shear stress is defined by [20]:

$$f_{vL} = f_{v,0} \frac{\emptyset e^{\emptyset}}{e^{\emptyset} - 1}$$
(19a)

where $f_{v,0}$ is the friction factor for single phase vapor flow defined for laminar flow as :

$$f_{v,0} = CRe_v^{-1}$$
(19b)

For turbulent flow, $f(v, 0)$ is determined from:

$$f_{v,0} = 0.0791 Re_v^{-0.25}$$
(20a)

$$Re_v = \frac{\rho_v U_v D_v}{\mu_v}, \quad D_v = \frac{4A_v}{P_{vL}},$$
(20b)

The factor $\varnothing$ is defined as the ratio of the local condensation mass flow rate to the vapor mass flow rate rebounding from the liquid-vapor interface [2], which is approximated by:

$$\varnothing = \frac{dq}{\mathcal{P}dz} \frac{2}{h_{fg} U_v \rho_v f_{v,0}}$$

(21)

Dimensionless Boundary Conditions

Equations (11) to (15) are solved using the following dimensionless boundary conditions:

1) At the microchannel inlet ($z^* = 0$):

- the flow mass flux G is imposed;
- the temperature and pressure at the saturated state are given;
- the non-dimensional vapor velocity is

$$U_v^* \left(z^* = 0 \right) = 1;$$

- the non-dimensional liquid velocity is

$$U_L^* \left(z^* = 0 \right) = 0 ;$$

- the non-dimensional curvature radius is

$$R^* \left(z^* = 0 \right) = \frac{\sigma}{P_{v0} D_h}$$

2) At the position $z^* = L^*$ corresponding to the end of the condensation zone:
- the non-dimensional outlet liquid pressure is:

$$P_l\left(z^* = L^*\right) = \frac{P_0 D}{2\sigma}$$;

- for the rectangular cross section, the dimensionless curvature radius is [22]:

$$R^*\left(z^* = L^*\right) = \varepsilon \sin(\beta) / (2\cos(\theta + \beta) D_h)$$

where $\varepsilon = \min(a, b)$;

- for the equilateral triangle cross section, the dimensionless curvature radius a is [18] :

$$R^*\left(z^* = L^*\right) = \mathcal{P} \sin(\beta) / (6\cos(\theta + \beta) D_h)$$

SOLUTION PROCEDURE

The above boundary conditions combined with nondimensional equations (11)-(15) constitute the mathematical model of two-phase flow in capillary regime with a vapor-liquid phase change. Solution of this mathematical model is not trivial since one of the limit positions remains to be found, the value of L being one of the unknowns of the problem. So, the iteration process is used to solve the mathematical model. To start the calculations, the saturated vapor at the inlet of the tube is assumed having the known mass flux, temperature, pressure, and vapor quality. For the next axial step $(z^* + \Delta z^*)$, the following steps are executed:

1) To start calculation, an arbitrary total condensation length is assumed;
2) Calculation of the curvature radius from equation (11);
3) Calculation of the local liquid velocity from equation (12);
4) Calculation of the local vapor velocity from equation (13);
5) Calculation of the liquid pressure from equation (14);
6) Calculation of the local vapor pressure from equation (15);
7) Calculation of the total condensation length from equation (26).

Steps 2) to 7) are repeated until the value of the condensation length obtained at the iteration number "it" is approximately equal to the one determined at iteration number "it-1".

Knowing the total condensation length, calculations are then made for the next values of z locations. For each z location, steps 2) to 7) are made and calculation is stopped when z value is approximately equal to the total condensation length L.

COMPUTATION OF THE HEAT TRANSFER COEFFICIENT AND THE CONDENSATION LENGTH

Calculation of the Heat Transfer Coefficient

Assuming that the temperature profile is linear in the condensate film, the local heat transfer is expressed by:

$$h(z) = \frac{\lambda_L}{\delta(z)}$$

(22a)

and hence the average heat transfer coefficient in the microchannel condensation length L is performed by the relationship:

$$h_m = \frac{1}{L} \int_0^L h(z) \, dz$$

(22b)

Two last quantities are used to determine the average and the local Nusselt numbers respectively:

$$Nu_{av} = \frac{h_m D}{\lambda_L}$$

(23a)

$$Nu_{loc} = \frac{h(z) D}{\lambda_L}$$

(23b)

$\delta(z)$ is the average condensate film thickness for each z location along the microchannel. Using the geometrical considerations from the Figure 3, the following analytical expression is established:

$$\delta(z) = \frac{1}{\pi} \int_0^{\frac{\pi}{2}} R \left[\frac{\cos(\theta)}{\cos(\Omega)} - 1 \right] d\Omega$$

(24)

5.2. Calculation of the Condensate Length

The parameter dR/dz given by the equation (9) is expressed in terms of the limit curvature radius R_L.

At $z = L$ location, dR/dz is to be infinite which causes a calculation problem. To avoid this complication, we set $z = L - \varepsilon$ where ε is an infinitesimal parameter. The limit values of the curvature radius and its first derivative are defined as [22]:

$$R_L = R\big|_{z=L-\varepsilon} = \frac{\varepsilon \sin \beta}{2\cos(\theta + \beta) D_h}$$

(25)

$$\frac{dR}{dz}\bigg|_{z=L-\varepsilon} = \frac{\varepsilon \cos\theta - R_L}{R_L \cos\theta}$$

(26)

To estimate the numerical value of the condensation length a dichotomy method was performed using equation (26), between the inlet of the microchannels which corresponds to ($z = 0$ and $R = 0$) and the end of the condensation zone where $(z = L - \varepsilon$ and $R = R_L)$.

RESULTS AND DISCUSSIONS

Numerical Results Validation

The validation of the results obtained in the present work for condensation of water in a square microchannel, are compared to the predictions of various correlations available in the literature. These correlations are proposed for condensation heat transfer in microchannels and macrochannels. Among these predictive correlations, those of condensation in microchannels are defined by Wang et al. [26], Koyama et al. [23], and Wang & Rose [14]. Figure 4 shows the comparisons between the average Nusselt number obtained from the present numerical model and those predicted by seven correlations: Dobson et al. [22], Wang et al. [26], Koyama et al. [23], Wang & Rose [14], Traviss et al. [24], Shah [19], and Ackers et al. [25]. Results presented in the Figure 4 are those obtained for condensation of steam in a horizontal square microchannel with hydraulic diameter of 110 μm, for various inletvapor mass fluxes

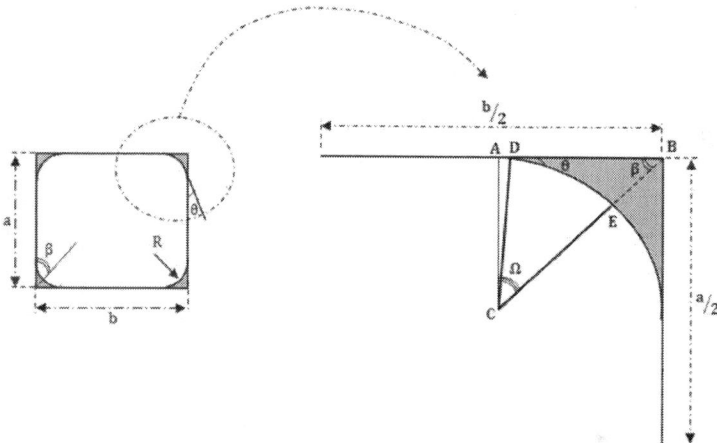

Figure 3: Distribution of the condensate film in the microchannel corners.

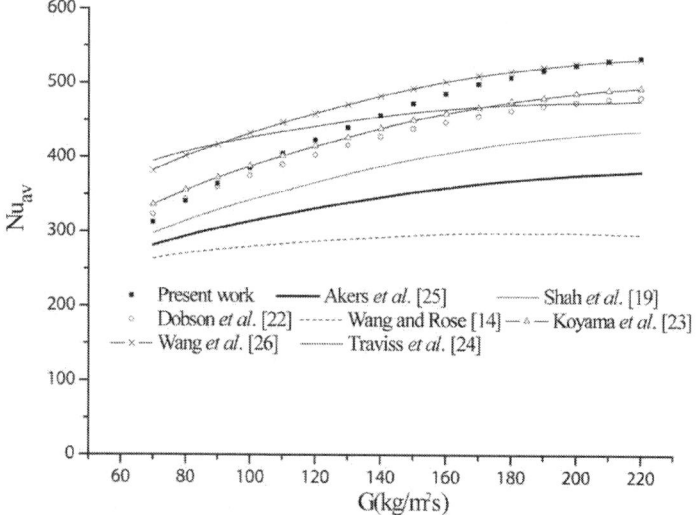

Figure 4: Comparison of the present work with predictions of literature correlations.

ranging from 70 to 220 kg/m²s. It is found that the best predictions of the present average Nusselt number are obtained by the correlations of Dobson et al. [22] and Koyama et al. [23] for which the mean relative deviation is about 7%. Both of these correlations are developed in the case of an annular condensation heat transfer of several pure refrigerants which is carried out in the channels with small hydraulic diameter. The correlations of Traviss et al. [24] and Ackers et al. [25] devoted to the convective condensation heat transfer in the macro-scale channels have been also evaluated to the micro-scale. It can be seen that the correlation of Traviss et al. [24] over predicts highly the present results for low mass fluxes whereas the Ackers's correlation [25] under predicts them highly for all the range of the tested mass fluxes. Adding up, the correlation of Shah [19] is proposed for condensation heat transfer in channels with large hydraulic diameter from 7 to 40 mm and Wang et al. [26] correlation is defined for condensation heat transfer inside a horizontal small rectangular channel with hydraulic diameter of 1.46mm. Average Nusselt numbers obtained by both of these correlations give the same trend and the reasonable results compared to those predicted numerically. The Shah's correlation under predicts numerical results with a maximum average deviation of about 16%. As for the correlation of Wang et al. [26], it over predicts the present results with a maximum deviation of 22% obtained at low mass fluxes. Moreover, correlation of Wang & Rose [14] based on the Nusselt theory including the effects of interfacial shear stress and surface tension on condensation heat transfer is also evaluated. Predictions of Nusselt number from the correlation of Wang & Rose [14] are not in accordance with those of this study since they gave the highest deviation of 37% form the numerical results.

Peripheral Condensate Film Thickness and Local Heat Transfer
Numerical results are given in the Figures 5(a) and (b) and the Figures 6(a) and (b) for steam condensation in square section microchannel with hydraulic diameter of 110 and 250 μm. Computations were conducted for vapor mass flux G = 90 kg/m²s, contact angle q = 15°, heat flux density q = 100 kW/m². For the conditions used in the present computation, the boiling number, the capillary number, the inlet steam temperature and pressure are maintained constant.

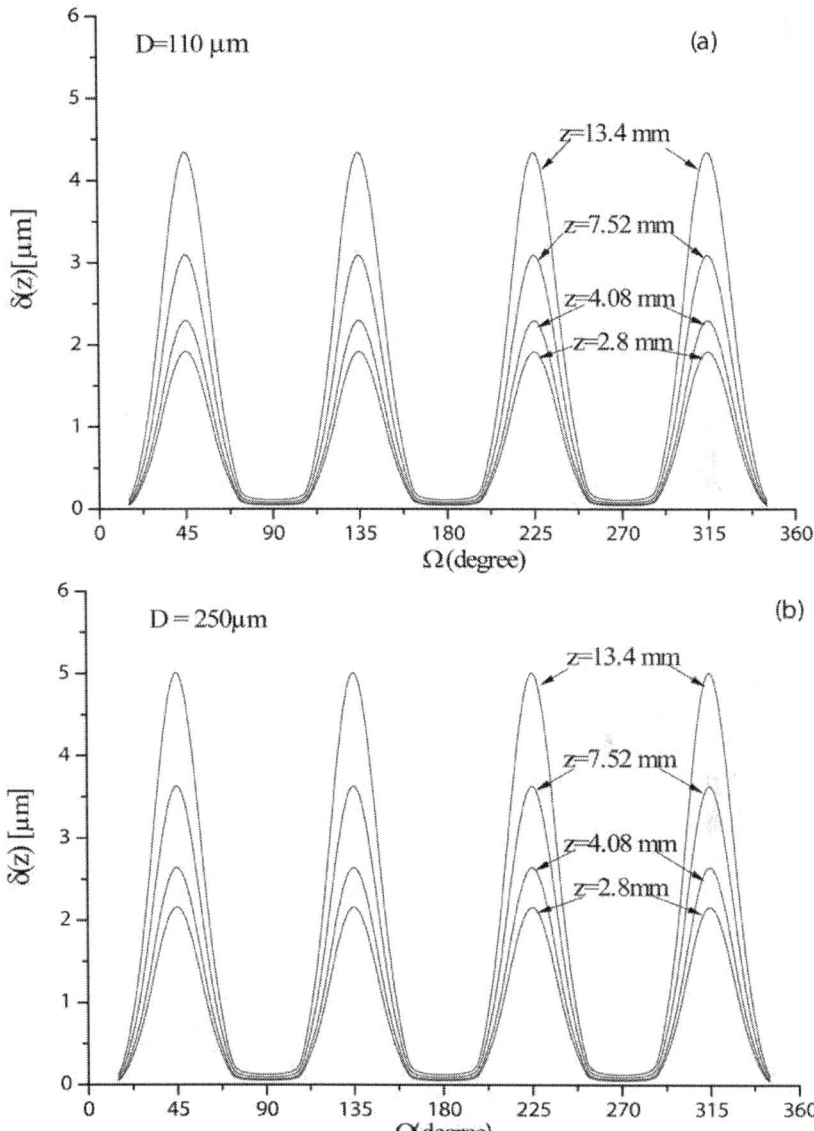

Figure 5: Evolution of the condensate film thickness around the square microchannel circumference.

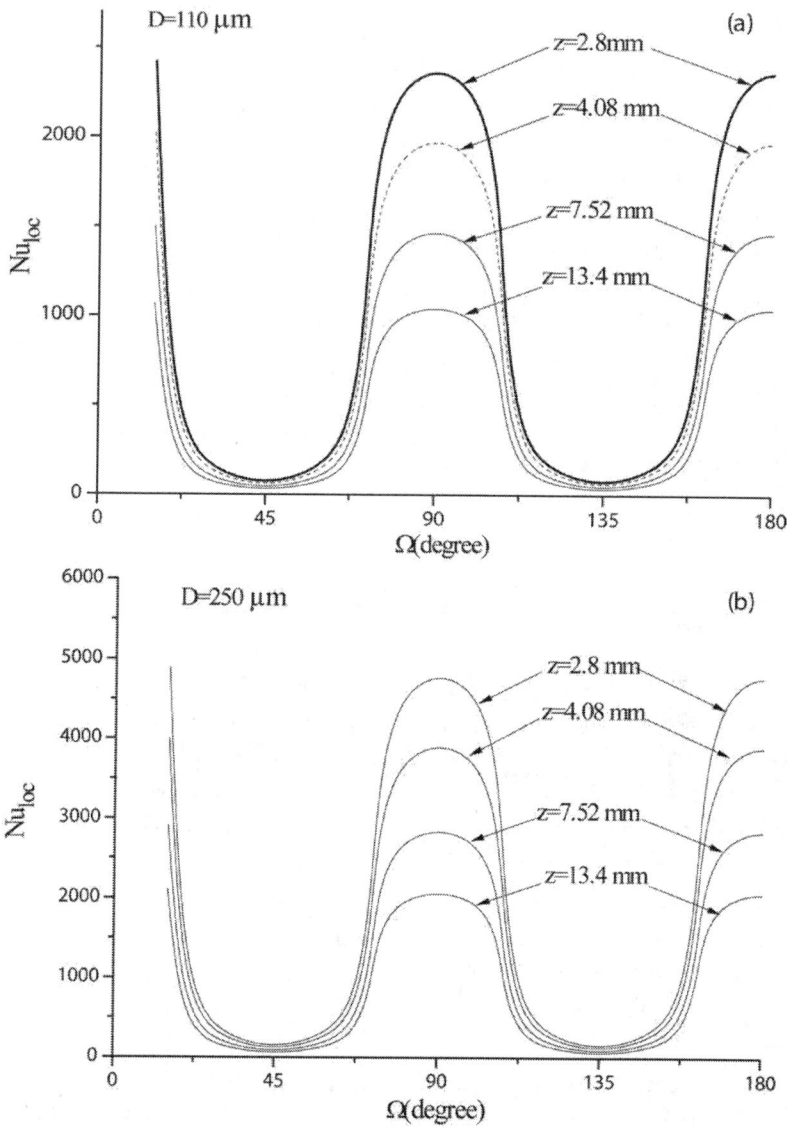

Figure 6: Evolution of the local Nusselt number around the square microchannel circumference.

To better understand the behavior of the peripheral local heat transfer coefficient, Figures 5(a) and (b) show predicted square channel condensate film thickness (24) plotted as a function of the peripheral coordinate W at four different axial locations. It can be seen from these figures that for the same z location on the microchannel, the condensate film thickness is higher for 250 μm (**Figure 5**(b)) than for 110 μm (**Figure 5**(a)). According to these representations, it is observed that for both microchannels, the symmetrical distributions of the condensate film thickness are obtained around the channel perimeter because the gravity has no effect on the condensate flow in microchannel. The condensate film thickness is very thick in the microchannel corners (W = 45°, 135°, 225° and 315°) under the effect of the surface tension whereas it becomes very thin at W = 0°, 90°, 180° and 360 where the heat transfer is the highest. These trends have been reversed for the peripheral local heat transfer coefficient (see Figures 6(a) and (b)) even if the symmetrical distributions are conserved.

Influence of the Cross-Section Shape on the Heat Transfer Coefficient

To indicate the influence of the channel cross-section shape, we start by the presentation of the results concerning curvature radius. In fact, Figure 7 shows the variation of the dimensionless curvature radius along the flow direction in three cross-section shapes of microchannels. At the beginning and the end of the condensation, the curvature radius increases rapidly along the channel, while it increases slowly at the middle. We also note that for the triangular microchannel it's increasing very quickly than other geometries. The present simulation results are similar to those of the reference [21].

The annular condensation length is one of the most important parameter which influences the thermal performance of the microchannel studied here. Figure 8 gives the dimensionless annular condensation length with respect to the boiling number Bo, for three cross-section shapes. As shown in this figure, the condensation zone decreases with Bo and it is also clear that the cross-section shape of the microchannel plays a noticeable role on the condensation.

In the heat transfer exchange point of view, the influence of the microchannel shape on the average Nusselt number is highlighted by studying condensation of steam in a square, equilateral triangular, and rectangular microchannels with the same hydraulic diameter of 250 μm. The sides of the equilateral triangular and square microchannels are 433 μm and 250 μm respectively. For rectangular microchannels having the same hydraulic diameter, the aspect ratio is about 2, 3 and 4. Three different rectangular cross sections are investigated: 375 × 187.5 μm², 500 × 166.6 μm² and 625 × 156.25 μm².

Under the same conditions, Figure 9 compares the average Nusselt numbers with respect to the vapor Reynolds number for five microchannel cross-sections. According to this figure, it is seen that annular condensation Nusselt number is low for square microchannel cross section (aspect ratio of 1). For rectangular cross-section microchannels with aspect ratio (b/a) higher than 1, condensation Nusselt number increases by increasing aspect ratio. The highest Nusselt numbers values are obtained

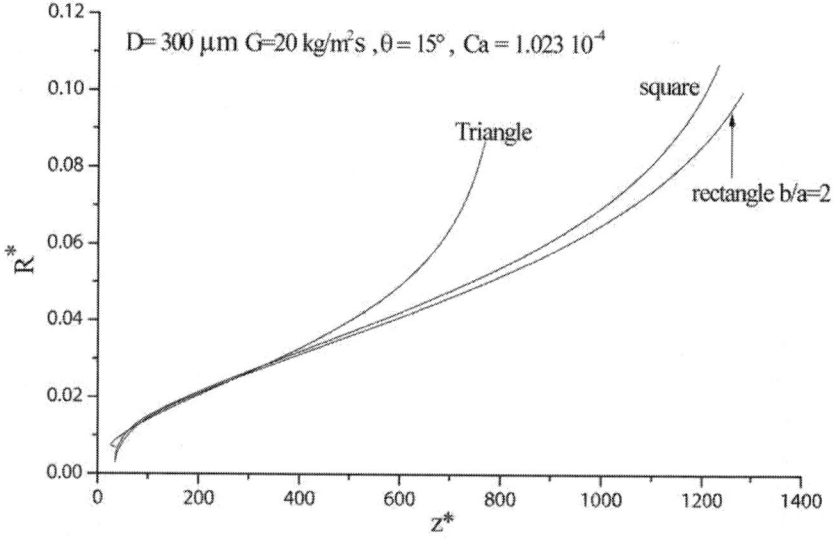

Figure 7: Dimensionless curvature radius along various microchannel shapes.

Vl: Liquid-vapor interface

O: Inlet

CITATION

Hicham El Mghari, Mohamed Asbik, Hasna Louahlia-Gualous, Condensation Heat Transfer in Horizontal Non-Circular Microchannels. Energy and Power Engineering Vol.5 No.9(2013), Article ID:40258,10 pagesDOI:10.4236/epe.2013.59063

Chapter 7

Analytical and Experimental Investigation About Heat Transfer of Hot-Wire Anemometry

Mojtaba Dehghan Manshadi and Mohammad Kazemi Esfeh

[1]Malekashtar University of Technology, IranUniversity of Yazd, Yazd, Iran

INTRODUCTION

The hot-wire anemometer is a famous thermal instrument for turbulence measurements [1]. The principle of operation of the anemometer is based on the heat transfer from a fine filament where it is exposed to an unknown flow that varies with deviation in the flow rate. The hot-wire filament is made from a special material that processes a temperature coefficient of resistance [2]. Thermal anemometry is the most common method employed to measure instantaneous fluid velocity. It may be operated in one of these two modes, constant current (CC) mode and constant temperature (CT) mode.

Constant-Current (CC) mode: In this mode, the current flow through the hot wire is kept constant and variation in the wire resistance caused by the fluid flow is measured by monitoring the voltage drop variations across the filament.

Constant Temperature (CT) mode: In this mode, the hotwire filament is positioned in a feedback circuit and tends to maintain the hotwire at a constant resistance and hence at a constant temperature and fluctuations in the cooling of the hot wire, filaments are similar to variations in the current flow through the hotwire.

Hot wire anemometers are normally operated in the constant (CTA) mode. The hot-wire anemometry has been used for many years in fluid mechanics as a relatively economical and effective method of measuring the flow velocity and turbulence. It is based on the convective heat transfer from a heated sensing element. Briefly; any fluid velocity change would cause a corresponding change of the convective heat loss to the surrounding fluid from an electrically heated sensing probe. The variation of heat loss from the thermal element can be interpreted as a measure of the fluid velocity changes. In subsonic incompressible flow the heat transfer from a hot wire sensor is dependent on the mass flow, ambient temperature and wire temperature. Since density variations are assumed to be zero, the mass flow variations are only function of velocity changes. The major advantage of maintaining the hot wire at a constant operational temperature and thereby at a constant resistance is that the thermal inertia of the sensing element is automatically adjusted when the flow conditions are varied. The electronic circuit of chosen CTA is shown schematically in Fig. 1. This mode of operation is achieved by incorporating a feedback differential amplifier into the hot-wire anemometer circuit. Such set-up obtains a rapid variation in the heating current and compensates for instantaneous changes in the flow velocity [2]. The sensing element in case studied in this research is a tungsten wire that is heated by an electric current to a temperature of approximately 250°C. The heat is transferred from the wire mainly through convection. This heat loss is strongly dependent on the excess temperature of the wire, the physical properties of the sensing element and its geometrical configuration. The authors strive to present an analytical solution for heat transfer equation of hotwire for states that can ignore the radiation term for the wire. The fundamental principle of hot-wire anemometer is based on the convective heat transfer, thus in the research, an attempt is made to develop a better perception from the heat transfer of the hot-wire sensor. Also, the effect of air flow temperature variations on the voltage of hot wire, CTA has been studied experimentally. Furthermore, on the basis of air flow velocity and ambient temperature variations, the percentage errors in velocity measurements have been calculated. Finally, based on results, an accurate method has been proposed to compensate for air flow temperature variations.

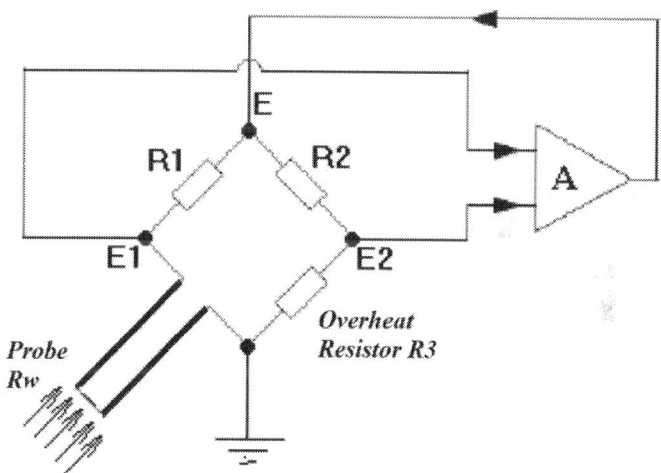

Figure 1: Schematic of a Constant Temperature Anemometer.

THEORETICAL BACKGROUND

The hot-wire involves one part of a Wheatstone bridge, where the wire resistance is kept constant over the bandwidth of the feedback loop. The electrical power dissipation Q'elec, when the sensor is heated, is given by:

$$\dot{Q}_{elec} = I^2 R_w \tag{1}$$

I and R_w are the current passing through the sensor and the resistance of the sensor at the temperature Tw, respectively. The convection heat transfer rate to the fluid can be expressed in terms of the heat-transfer coefficient h, as:

$$\dot{Q} = A_w h \left(T_w - T_a \right) \tag{2}$$

Where A_w is the surface area of sensor and (T_w-T_a) is the difference between the temperature of the hot-wire sensor and the temperature of the fluid. For steady-state operation, the rate of electric power dissipation equals to the rate of convective heat transfer (assuming the conductive heat transfer to the two prongs is negligible). Thus,

$$I^2 R_w = \pi d L h \left(T_w - T_a \right) = \pi L k \left(T_w - T_a \right) Nu \tag{3}$$

By introducing the wire voltage Ew = IRw and using equation (3), one can conclude that (k is the thermal conductivity of the fluid):

$$\frac{E_w^2}{R_w} = \pi L k \left(T_w - T_a \right) Nu \tag{4}$$

According to the pioneering experimental and theoretical work by King, the convective heat transfer is often expressed in the following form:

$$Nu = A + BRe^{n} \tag{5}$$

Where A and B are empirical calibration constants. For long wires in air, King found that A=0.338, B=0.69 and n=0.5. It is interesting to note that King based his derivations on the assumption of potential flow, which is a poor approximation of the real flow around a wire at low Reynolds' numbers, so King's derivation is in a sense approximately erroneous. Nevertheless, King's law has been the considered tool for fitting calibration data in practical hot-wire anemometry for almost a hundred years [3].

By introducing equation (5) into equation (4) can give:

$$\frac{E_w^2}{R_w} = \pi L k \left(T_w - T_a \right) \left(A + B \left(\frac{\rho d}{\mu} \right)^{0.5} U^{0.5} \right) \tag{6}$$

Equation (6) states that the hot-wire voltage is sensitive both to the velocity and temperature of air. Here, rearranging the equation (6) gives:

$$\frac{E_w^2}{R_w} = (A + BU^{0.5})(T_w - T_a) \tag{7}$$

Where π, l, k, d, ρ and μ have been included in the constant coefficients A and B.

According to equation (7), Kanevce and Oka [4] introduced the following expression to correct the hot-wire output voltage for the temperature drift:

$$E_{corr} = E_w \left(\frac{T_w - T_a}{T_w - T_{a,r}} \right)^{0.5} \tag{8}$$

$T_{a,r}$ is ambient reference temperature during sensor calibration and T_a is ambient temperature during data acquisition where E_{corr} is corrected voltage. For a hot wire probe with a finite length active wire element, the conductive heat transfer to prongs must be taken into account. In practice this is often achieved by the modifying equation [7] as:

$$\frac{E_w^2}{R_w} = \left(A + BU^n \right)\left(T_w - T_a \right)$$

(9)

The values of A, B and n can be determined by a suitable calibration procedure. It should be noted that the term $(T_w - T_a)$ and physical properties of fluid are dependent on the ambient temperature. In the previous related studies, the effect of term $(T_w - T_a)$ is considered only to compensate the ambient temperature variations [5]. In other word, the variations of physical properties of fluid and Nusselt number are ignored. So in this study, the variations of Nusselt number with the fluid temperature have been considered. The following equation for correction of output voltage E has been proposed by the relations extended in Ref. [6].

$$E_{corr} = E_w \left(\frac{T_w - T_{a,r}}{T_w - T_a} \right)^{0.5(1 \pm m)}$$

(10)

In Ref. [7], equation (10) is employed to correct the voltage of CTA output. Results showed that the required error correction factor (m) depends on whether the fluid temperature decreases or increases with respect to the calibration temperature of the CTA.

The CT mode velocity and temperature sensitivities corresponding to equation (9) are:

$$S_u = \frac{\partial E_w}{\partial U} = \frac{nBU^{n-1}}{2} \left[\frac{R_w \left(T_w - T_a \right)}{A + BU^n} \right]^{0.5} \tag{11}$$

$$S_\theta = \frac{\partial E_w}{\partial \theta} = \frac{-1}{2} \left[\frac{R_w \left(A - BU^n \right)}{T_w - T_a} \right]^{0.5} \tag{12}$$

Where θ is a small fluctuation in the fluid temperature. Equations (11) and (12) show that the value of S_u increases and the value of S_θ decreases by increasing the value of $(T_w\text{-}T_a)$. A high over-heat ratio (R_w/R_a) is recommended for the measurement of velocity fluctuations [2]. In Ref. [7], it is stated that for an over-heat ratio of 1.4, the error incurred amounts to about 2.5% per degree Celsius temperature change. With the increase in the overheat ratio to 1.6, the error in CTA output is reduced to about 2% per degree Celsius temperature change.

The heat transfer process from a hot-wire sensor is usually expressed in a non-dimensional form where involve a relationship between the Nusselt number, the Reynolds number and the Prandtl number. The Nusselt number is usually assumed to be a function of Reynolds and Prandtl numbers and under most flow conditions, the Prandtl number is constant.

In hot-wire anemometry, the sizes of the sensing element are small, so that the Reynolds number of the flow is very low and the flow pattern over the sensor can be assumed to be symmetrical and quasi-steady. Due to the statement of the flow continuity, the mean free path of the particles is very much less than the diameter of the wire and conventional heat transfer theories are applicable [8]. Furthermore, the length of the sensor is much greater than its diameter. Hence, it may be assumed that the loss conduction through the ends is negligible and the relation for the heat transfer from an infinite cylinder can be applied. Kramers [9] has proposed the following equation based on heat-transfer experimental results for wires (with infinite length-to-diameter ratio), placed in air, water and oil:

$$Nu = 0.42 Pr^{0.2} + 0.57 Pr^{0.33} Re^{0.5} \qquad (13)$$

He selected the film temperature $T_f = (T_w + T_a)/2$ as the reference temperature for the fluid properties.

EXPERIMENTAL PROCEDURE

An air condition unit was used to carry out the experiments (Fig. 2). A laminar airflow was achieved by means of honeycombs network and screens. The air condition unit is powered by a small variable speed electric fan and four controllable heating elements provide a stable air temperature. The air flow velocity was measured by a pitot tube and a pressure transducer during the calibration and test. The output voltage from the hot-wire, pressure transducer output voltage, and the thermometer (NTC) output voltage are transferred to a computer, via an A/D card, having a 12 bit resolution and up to 100 kHz frequency.

The sensing element in our case is a standard 5μm diameter tungsten wire that is heated by an electric current to a temperature of approximately 250 °C. The active wire length is 1.25 mm. For such probes, the convective heat transfer is about 85 percent of the total heat transfer from the heated-wire element [2].

Before measurements, the hot-wire sensor was calibrated in a wind tunnel and the response of the anemometer bridge voltage was also expanded as a least square fit with a 5th order polynomial ($U = C_0 + C_1 E + C_2 E^2 + C_3 E^3 + C_4 E^4 + C_5 E^5$).

The experiments were carried out on a hot wire sensor operating at an over-heat ratio (R_w / R_a) of 1.8. The sensor, after calibration, was tested at different temperatures. The velocity range was 1-2 m/s, which corresponds to a Reynolds number of 0.18-0.35, and the temperature range, was 17.5-40 °C.

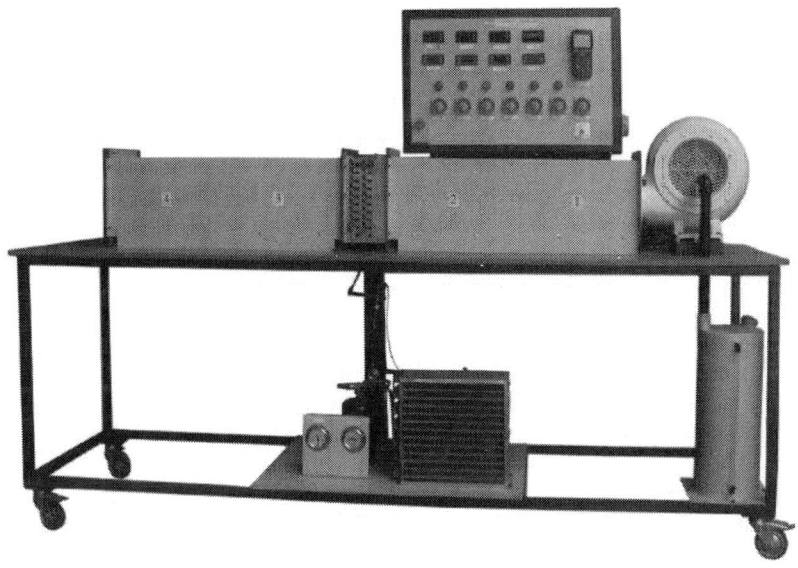

Figure 2: The Laboratory Air Condition Unit.

RESULTS AND DISCUSSIONS

To examine the behavior of the hot wire sensor in different conditions and determine the temperature distribution along it, the general hot wire equation must be derived initially. By considering an incremental element of the hot wire, Fig.3, an energy balance can be performed where assume that there is the uniform temperature over its cross-section according to the equation (14).

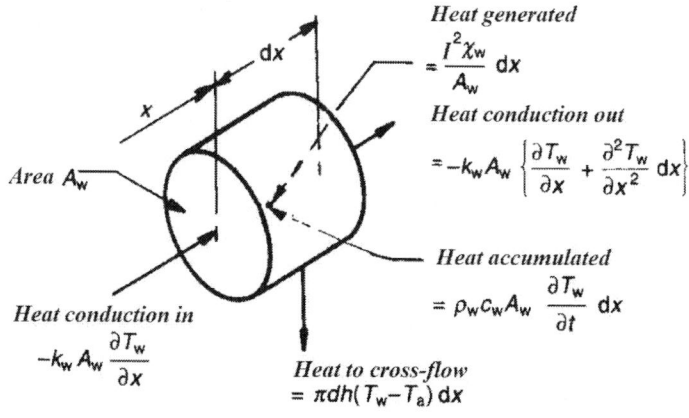

Figure 3: Heat Balance for an Incremental Element [2].

$$\frac{I^2 \chi_w}{A_w} = \pi dh(T_w - T_a) - K_w A_w \frac{\partial^2 T_w}{\partial x^2} + \pi d\sigma\varepsilon\left(T_w^4 - T_s^4\right) + \rho_w c_w A_w \frac{\partial T_w}{\partial t} \qquad (14)$$

Where I is electrical current, χ_w is the electrical resistant of the wire material at the local wire temperature, T_w, and A_w is the cross-sectional area of wire where h is the heat-transfer coefficient, c_w is the specific heat of the wire material per unit mass, k_w is the thermal conductivity of the wire material and d is the diameter of wire. With using the fourth-order Runge-Kutta method, this nonlinear secondary differential equation is solved in two conditions: with radiation term and without radiation term. Fig.4 shows the results for this step. As it is shown, the radiation term does not have any effect on the temperature distribution. The previous results achieved in Ref. [3] indicate that, the error due to radiation is in the range 0.1-0.01% and is quite negligible.

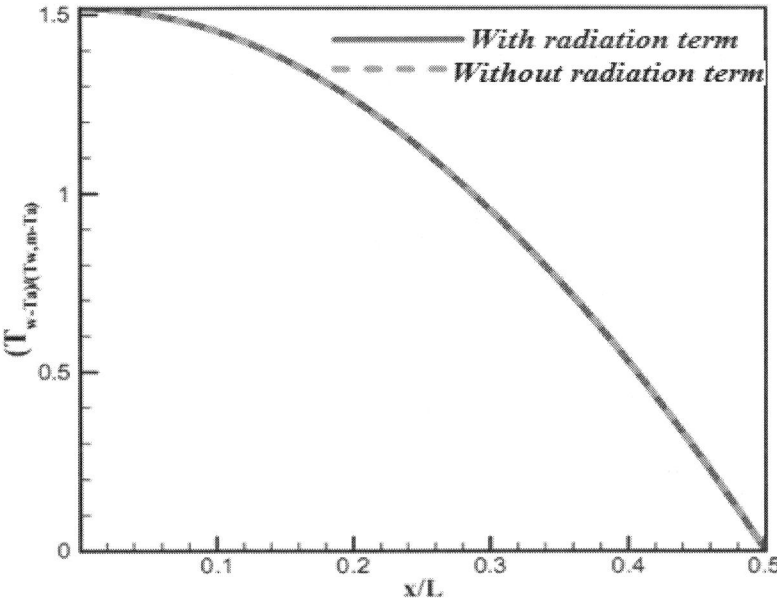

Figure 4: The Solution of Equation (14) With and Without Radiation Term.

Under steady conditions,

$$\frac{\partial T_w}{\partial t} = 0$$

Xw can be expressed as $\chi w = \chi a + \chi o \ \alpha_o \ (T_w - T_a)$. Where χa and χo are the values of the resistivity at the ambient fluid temperature, T_a, and at 0°C and α_o is temperature coefficient of resistivity at 0°C. Thus, equation (14) can be rewritten as equation 15 [2]:

$$K_w A_w \frac{d^2 T_w}{dx^2} + \left(\frac{I^2 \chi_o \alpha_o}{A_w} - \pi dh \right)(T_w - T_a) + \left(\frac{I^2 \chi_a}{A_w} \right) = 0 \qquad (15)$$

With assuming the ambient temperature is constant along the wire, this equation is of the following form (16):

$$\frac{d^2 T_w}{dx^2} + K_1 T_1 + K_2 = 0 \qquad\qquad (16)$$

Where

$T_1 = T_w - T_a$

$$T_1 = T_w - T_a$$

and

$$K_1 = \left(\frac{I^2 \chi_o \alpha_o}{A_w} - \pi dh \right)$$

$$K_2 = \left(\frac{I^2 \chi_a}{A_w} \right)$$

The value of K_1 may be negative or positive. Therefore, the solution of equation (16) and temperature distribution along the wire are dependent on the value of K_1. Equation (16) is solved in three states: $K_1 < 0$, $K_1 = 0$, $K_1 > 0$.

STATE I: $K_1 < 0$

For more hot-wire applications, K_1 will be negative [2]. In Ref. [2], it is declared that in this state, the solution for a wire of length L will become:

$$T_w = \frac{K_2}{|K_1|}\left[1 - \frac{\cosh\left(|K_1|^{0.5}x\right)}{\cosh\left(\frac{|K_1|^{0.5}L}{2}\right)}\right] + T_a \tag{17}$$

The mean wire temperature, $T_{w,m}$ is obtained by integrating equation (17):

$$T_{w,m} = \frac{1}{L}\int_{-L/2}^{L/2}T_w(x)dx \tag{18}$$

Inserting equation (17) into equation (18) gives:

$$T_{w,m} = \frac{K_2}{|K_1|}\left[1 - \frac{\tanh\left(|K_1|^{0.5}L/2\right)}{|K_1|^{0.5}L/2}\right] + T_a \tag{19}$$

The non-dimensional steady state wire temperature distribution will be achieved such as equation (20) [2]:

$$\frac{T_w - T_a}{T_{w,m} - T_a} = \frac{\left[\frac{L\times|K_1|^{0.5}}{2}\right]\left[\cosh\left(\frac{L\times|K_1|^{0.5}}{2}\right) - \cosh\left(x\times|K_1|^{0.5}\right)\right]}{\left[\frac{L\times|K_1|^{0.5}}{2}\cosh\left(\frac{L\times|K_1|^{0.5}}{2}\right) - \sinh\left(\frac{L\times|K_1|^{0.5}}{2}\right)\right]} \tag{20}$$

Where T_a is the ambient fluid temperature and $T_{w,m}$ is the mean wire temperature.

The convective and conductive heat transfer rate can be found from the flow conditions and the wire temperature distribution will earn according to the following equations:

$$\dot{Q}_{cond} = 2k_w A_w \left| \frac{dT_w}{dx} \right|_{x=l/2} \tag{21}$$

$$\dot{Q}_{conv} = \pi dh L (T_{w,m} - T_a) \tag{22}$$

To achieve a reasonable accuracy, the ratio of conductive heat transfer to convective heat transfer should be as low as possible.

$$\frac{\dot{Q}_{cond}}{\dot{Q}_{conv}} = 2 \frac{k_w A_w |K_1|^{0.5}}{\pi d L h} \times \frac{\tanh\left(0.5L|K_1|^{0.5}\right)}{\left(1 - \frac{2}{L|K_1|^{0.5}} \tanh\left(0.5L|K_1|^{0.5}\right)\right)} \tag{23}$$

According to equation (23), to reduce the effect of the conductive heat transfer rate, the wire should be as long as possible and the thermal conductive of the wire material should have a low value.

The temperature distribution in the form $(T_w - T_a)/(T_{w,m} - T_a)$, is shown in Fig.5. It is shown that the uniformity of the temperature distribution along the wire increases for longer length wires. Also, the value of temperature in different parts of the wire approaches to mean temperature with escalating the length wire.

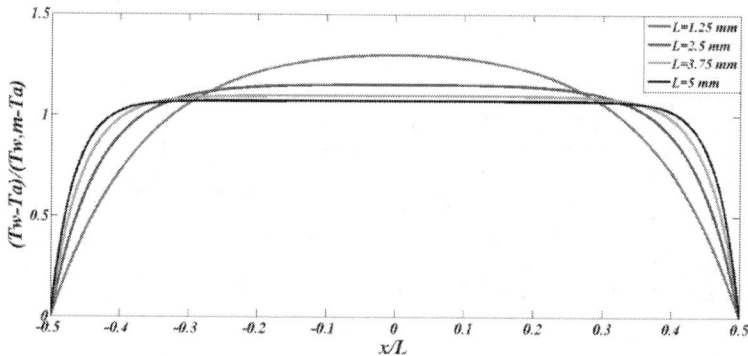

Figure 5: The Temperature Distribution along A Hot Wire for Various Values of L ($K_1<0$).

The effect of wire length on the percent of conduction and convection heat transfer is shown in Fig. 6. (Diameter of wire is 5μm and the air velocity is equal to 20 m/s). As it is shown the conductive end losses reduces with increasing the wire length but it should be noted, the maximum value of tanh (0.5L $|K_1|^{0.5}$) is approximately 1, so exceeding the wire length over 5.3/$|K_1|^{0.5}$ will not cause a reduction in the conductive heat transfer rate.

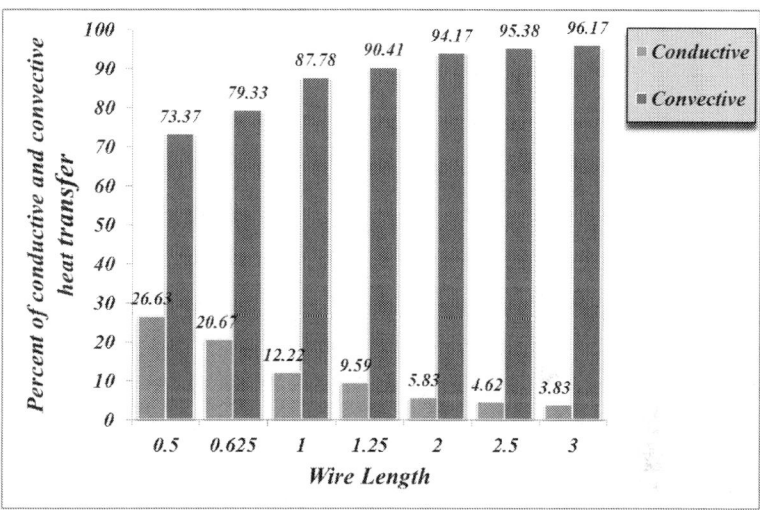

Figure 6: The Percent of Conduction and Convection Heat Transfer for Different Wire Length.

For hot-wire anemometer applications it is usually advantageous to minimize the rate of conductive heat transfer rate relative to the forced convective heat transfer rate [2]. Fig.7 shows the effect of wire diameter on the non-dimensional temperature distribution. It is shown, the uniformity of temperature distribution decreases with increasing the wire diameter. This variation is due to increasing the wire diameter that will cause the conductive heat transfer rate to the two prongs to be increased as well.

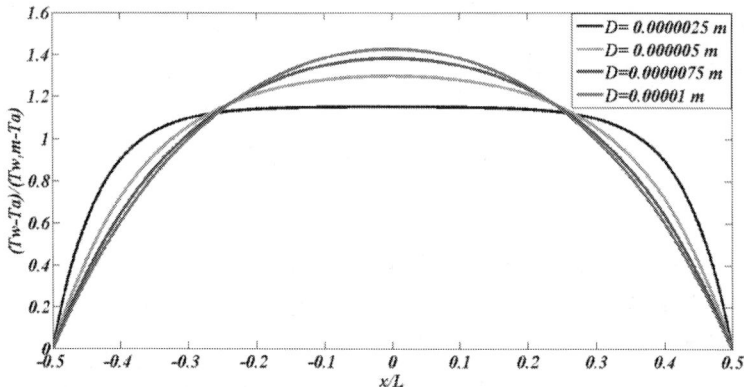

Figure 7: The Temperature Distribution along the Hot Wire for Various Values of D (K1<0).

Comparing between Figs. 6 and 7 shows that the wire diameter has the greater influence on the temperature distribution rather than the wire length.

State Ii; $K_1=0$

In this state, the temperature distribution equation is obtained as:

$$T_w\left(x\right) = -\frac{1}{2}K_2L^2\left[\left(\frac{x}{L}\right)^2 - \frac{1}{4}\right] + T_a \tag{24}$$

According to equations (24) and (18), the mean wire temperature is determined as:

$$T_{w,m} = \frac{1}{12}K_2L^2 + T_a \tag{25}$$

By using equations (24) and (25), the non-dimensional wire temperature distribution can be expressed as:

$$\frac{T_w(x) - T_a}{T_m(x) - T_a} = -6\left[\left(\frac{x}{L}\right)^2 - \frac{1}{4}\right] \qquad (26)$$

For this state, the non-dimensional temperature distribution is shown in Fig.8. It can be observed that the temperature distribution along the wire is independent of the wire length and for various values of L, all temperature profiles are identical. Also, it can be observed from equation (26) that the non-dimensional temperature distribution does not depend on the wire diameter.

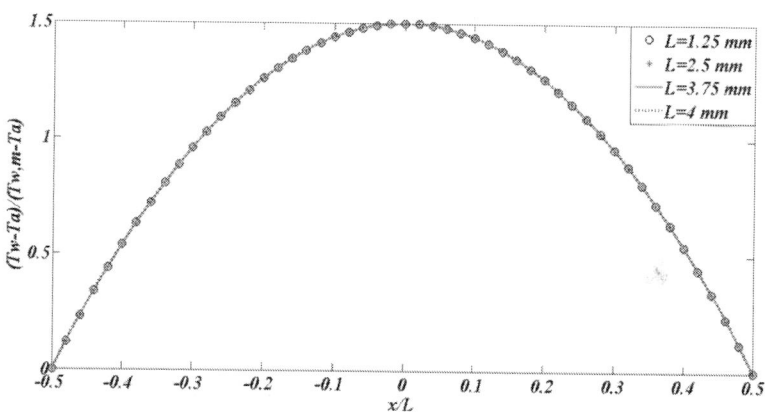

Figure 8: The Temperature Distribution along A Hot Wire for Various Values Of L (K1=0).

Furthermore, the ratio of conductive heat transfer rate to the forced convective heat transfer rate can be expressed as:

$$\frac{\dot{Q}_{cond}}{\dot{Q}_{conv}} = \frac{3}{2}\frac{k_w d}{hL} \qquad (27)$$

While it is shown, this ratio is directly proportional to the length and it will increase with inverse proportion to the wire diameter.

STATE III; $K_1 > 0$

Using the mathematical analysis, it can be demonstrated that temperature distribution equation is:

$$T_w = \frac{K_2}{K_1}\left[\frac{\cos\left(\left(K_1\right)^{0.5} x\right)}{\cos\left(\frac{\left(K_1\right)^{0.5} L}{2}\right)} - 1\right] + T_a \tag{28}$$

The mean wire temperature and the non-dimensional wire temperature distribution can be expressed as:

$$T_{w,m} = \frac{K_2}{K_1}\left[\frac{\tan\left(\left(K_1\right)^{0.5} L/2\right)}{\left(K_1\right)^{0.5} L/2} - 1\right] + T_a \tag{29}$$

$$\frac{T_w - T_a}{T_{w,m} - T_a} = \frac{\left[\frac{L\times\left(K_1\right)^{0.5}}{2}\right]\left[\cos\left(x\times\left(K_1\right)^{0.5}\right) - \cos\left(\frac{L\times\left(K_1\right)^{0.5}}{2}\right)\right]}{\sin\left(\frac{L\times\left(K_1\right)^{0.5}}{2}\right) - \left[\frac{L\times\left(K_1\right)^{0.5}}{2}\right]\cos\left(\frac{L\times|K_1|^{0.5}}{2}\right)} \tag{30}$$

Fig.9 shows the non-dimensional temperature distribution ($K_1 > 0$). As it is shown, in this state some fluctuations appear in the temperature profiles. It can be demonstrated from equation (28) that, with approaching the wire length to $\pi/|k_1|^{0.5}$, these fluctuations increases wherever the amplitude oscillatin decreases with growing the wire length.

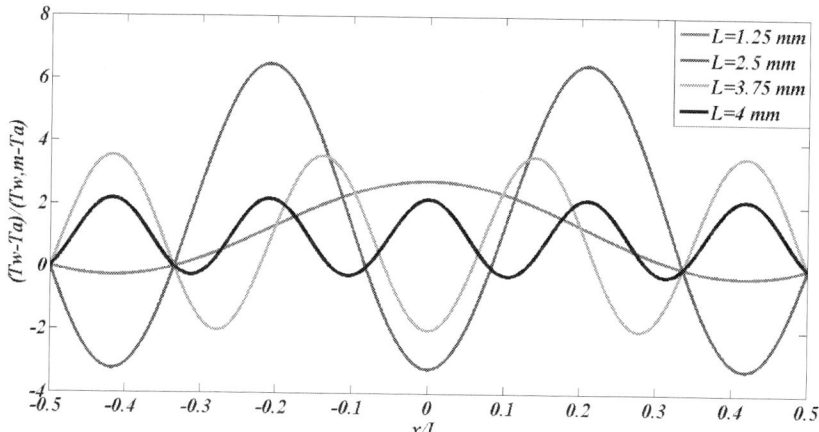

Figure 9: The Temperature Distribution along A Hot Wire for Various Values of L (K1>0).

It should be noted that the temperature profile is strongly dependent on the value of K_1 which is relevant to the heat transfer coefficient (Nusselt number). With setting the value of K_1 to zero, one could determine the critical Nusselt number as the following equation:

$$Nu_{critical} = \frac{I^2 \chi_o \alpha_o}{A_w \pi k} \qquad (31)$$

In summary, the authors consider the temperature distribution along the hot-wire in the following cases:

Case I, Nu>Nu critical: in this case, increasing the wire length and decreasing the wire diameter will cause the uniformity of temperature distribution to be increased considerably.

Case II, Nu=Nu critical: in this case, the temperature distribution is independent of length and diameter of wire.

Case III, Nu<Nu critical: here, temperature distribution is non-uniform and there are some fluctuations in temperature distribution.

According to equation (4), by knowing E_w, R_w, T_w and T_a in the anemometer, one can calculate the Nu number. The electrical resistance of the wire's material increases linearly with temperature, so that the resistance can be described as:

$$R_w = R_o\left[1 + \alpha_o(T_w - T_o)\right] \tag{32}$$

Where R_0 is the value of the resistance at a reference temperature T_0 and α is the temperature coefficient of resistance. The recommended value for over-heat ratio is equal to 1.8 and the wire temperature of the chosen probe is then 249.22 ℃.

In practical application, the hot-wire anemometer output is bridge voltage E (Fig. 1) whereas for determining the Nusselt number, the value of E_w is required. For a balanced anemometer bridge, the relationship between E (bridge voltage) and E_w (hot-wire sensor voltage) is:

$$E_w = \frac{E}{R_1 + R_w} R_w \tag{33}$$

For comparison, the calculated Nusselt number by equation (4) that it is based on the fluid properties evaluated at the film temperature defined as the mean of the upstream flow temperature and temperature on the hot wire versus Reynolds number where based on the fluid properties evaluated at the film temperature is presented in Fig.10 with Kramer's experimental formula.

As it is shown, the data does not collapse to one curve and the deviation increases with increasing Reynolds number. Our results are lower than those given by Kramer's formula and the differences may be caused by the effect of conductive heat transfer to the prongs and three-dimensional effect encountered in experiments. In Ref. [2], it is stated that for a standard probe (d=5μm and l=1.25 mm), the conductive heat transfer to the two prongs is about 15 percent of the total heat transfer from the heated-wire element. Although, the results in this study show that at the high velocity, the percentage of error between the predicted Nusselt number by Kramer's formula and the achieved Nusselt number in this study is 50%. This result confirms that, there is a significant difference between the heat transfer process from finite length hot-wire sensor and infinite length one.

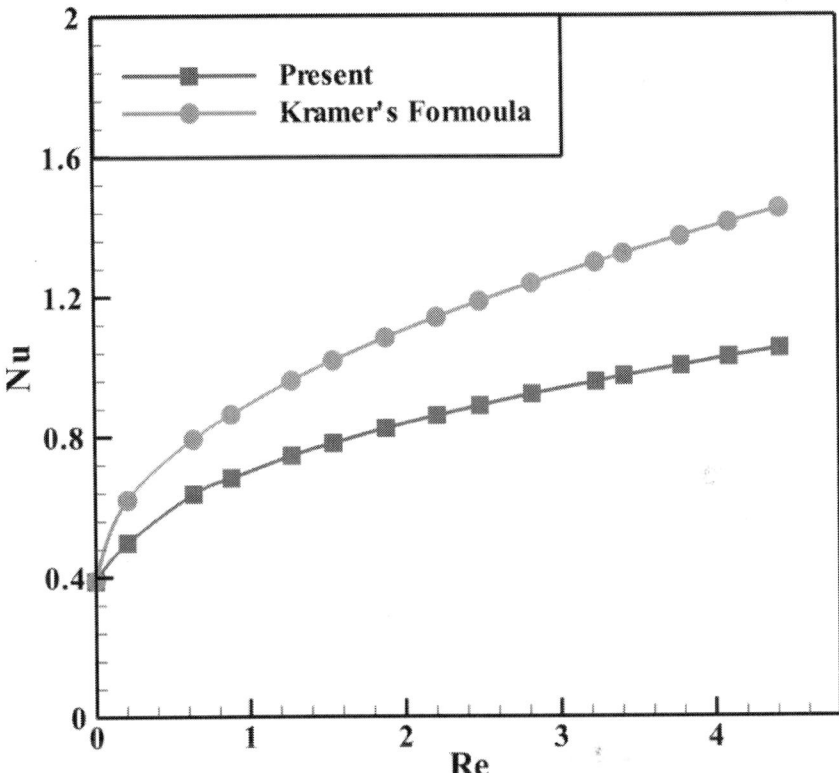

Figure 10: Heat Transfer from a Hot Wire in a Free Stream Flow.

In the next phase, the thermal response of the hot-wire anemometer relative to the velocity and also the air ambient temperature variation is investigated [11]. Figs. 11 and 12 show the variation of convection heat transfer coefficient (h) and Nusselt number versus Reynolds number at different ambient temperatures where fluid properties evaluated at the film temperature $T_f = (T_a+T_w)/2$.

The achieved results indicate convection heat transfer coefficient and Nusselt number vary with variation of air flow temperature and as expected, both of them decrease with increasing the ambient temperature.

Different equations have been proposed to modify the Nusselt number. Lundström et al. [3] claim that it was necessary to evaluate the fluid properties at the air temperature and their results show that evaluating the properties at the film temperature is not enough to achieve a temperature independent calibration law. Collis and Williams [10] realized, using the film reference temperature, that it was necessary to include a temperature loading factor in the Nusselt number King's law according to the following equation;

$$Nu\left(\frac{T_f}{T_a}\right)^{-0.17} = A + BRe^{0.45} \tag{34}$$

Nusselt number (hd/k) includes both the heat-transfer coefficient and the thermal conductivity of the fluid and these parameters are dependent on the ambient temperature. The temperature role on k can be expressed as:

$$\frac{k}{k_r} = \left(\frac{T}{T_r}\right)^{a} \tag{35}$$

T and T_r are in absolute temperature but the variation of h with ambient temperature is unknown.

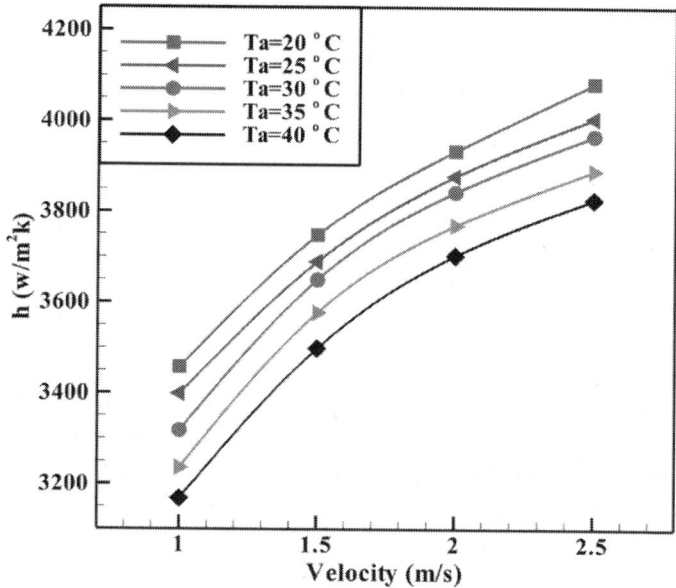

Figure 11: Variations of Heat-Transfer Coefficient vs. Velocity at Different Ambient Temperatures. [11]

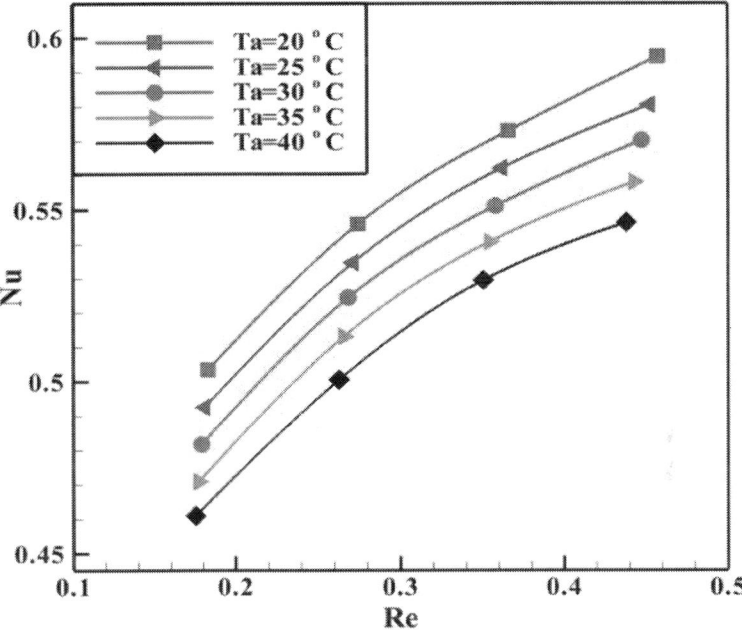

Figure 12: Variations of Nusselt Number vs. Re at Different Ambient Temperatures. [11]

However, the fundamental mechanism for variation of Nusselt number is not known yet but it can be compensated empirically by introducing the modified Nusselt number according to the following equation [11]:

$$Nu_{corr} = Nu\left(\frac{T_a}{T_{a,r}}\right) \tag{36}$$

Where temperatures are in absolute temperature and $T_{a,r}$ is the reference temperature at which the sensor calibration is performed. When the correction is applied to the data in Fig. 12, the data collapse approximately to a single curve as shown in Fig.13.

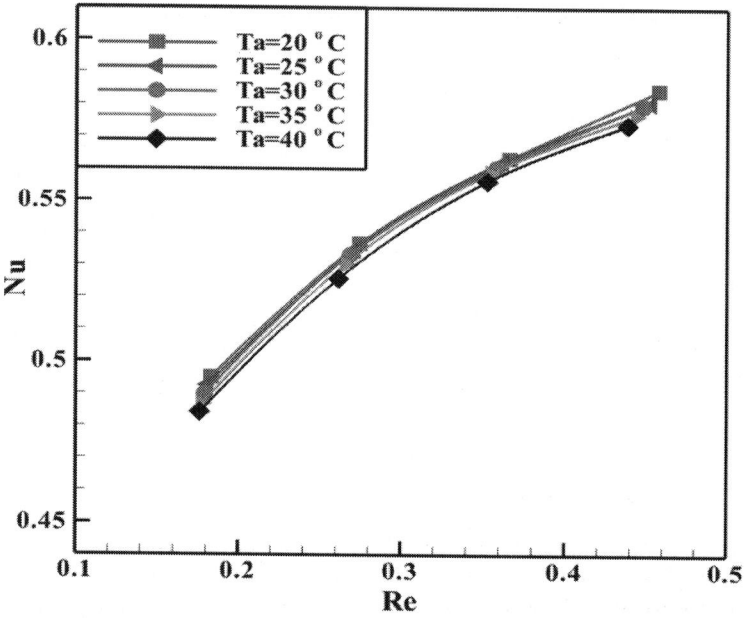

Figure 13: The Achieved Nusselt Number Vs. Re By Employing Equation (36) [11].

Fig. 14 presents the response of the CTA on wind speed approximately equal to 1 and 2 m/s at various temperatures. The temperature varies between 22.5 and 37.5°C. As it is shown, the bridge voltage decreases as the higher ambient temperature.

Calibration equations do not include ambient temperature variations, so a correction procedure should be applied. There are three main practical ways [2]:

1) Automatic compensation: Use a temperature sensor in the Wheatstone bridge.

2) Manual adjustment: Manual adjustment can be made by changing the value of the resistant, R_w, to compensate the changes in T_a.

3) Analytical correction: Measure the flow temperature separately and compensate using the heat transfer equation.

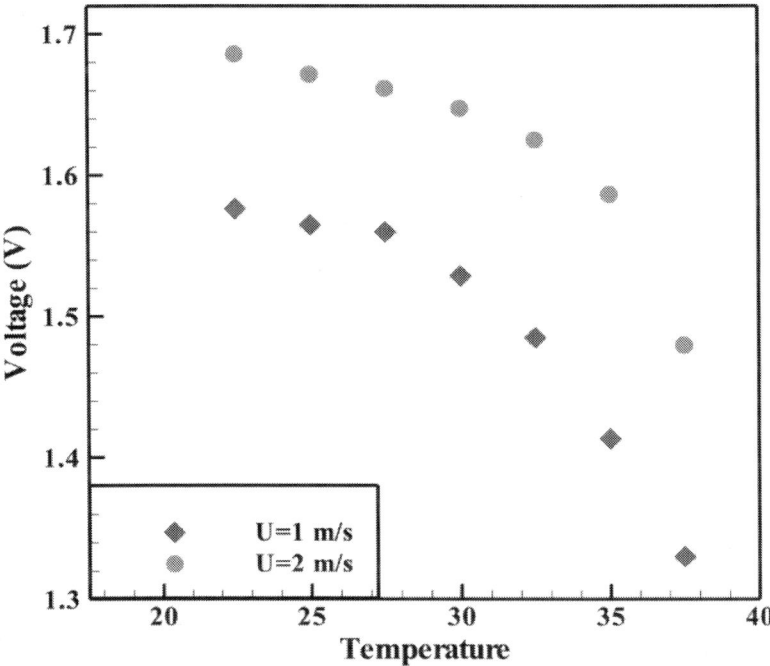

Figure 14: Response of cta at Various Ambient Temperatures. [11]

In this research, the voltage error due to changes in the ambient temperature is corrected using equation (10) and (37):

$$E_w = E\left(\frac{T_w - T_a}{T_w - T_{a,r}}\right)^{0.5} \times \left(\frac{T_a}{T_{a,r}}\right)^{0.5} \tag{37}$$

Equation (10) only considers the effect of ambient temperature variation but equation (37) regards the effect of Nusselt variation as well the ambient temperature variation.

The percentage error is presented in Fig.15 as a function of flow temperature. The other parameters are reference temperature, $T_{a,r}$=25 °C, average sensor temperature, T_w=249.22°C and flow velocity, U=1.5 m/s.

It can be seen that the achieved results from equation (37) are more reasonable. It should be noted that by increasing the air temperature, the fluid properties will be changed and these changes have to be taken into consideration. This factor is considered in equation (37) so that at high air temperature it can compensate the ambient temperature variations adequately.

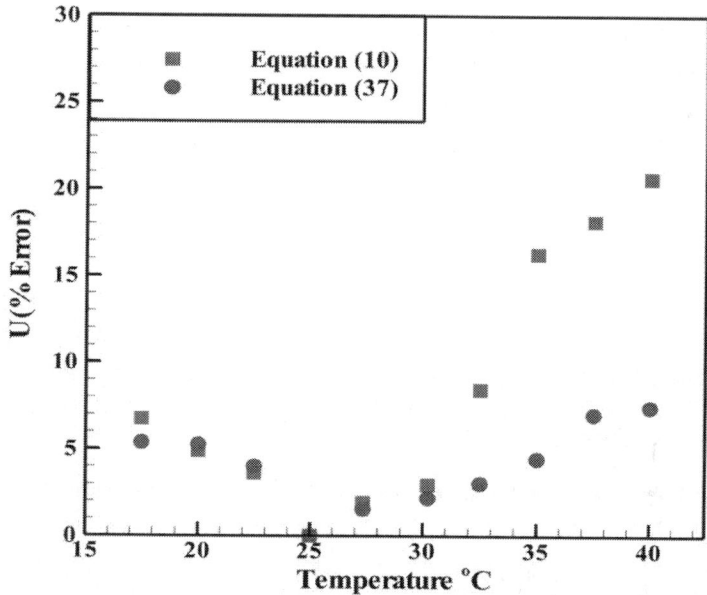

Figure 15: Error in the Measurement of Flow Velocity for the Hot Wire Sensor [11]

CONCLUSION

The analytical solutions for heat transfer equation of hotwire indicate that the temperature distribution along a hot-wire sensor is dependent on the critical Nusselt number. $Nu<Nu_{critical}$ leads to increasing the wire length and decreasing the wire diameter which will cause the uniformity of temperature distribution to be increased considerably. For $Nu=Nu_{critical}$, wire length and diameter don't have any effect on the temperature distribution. If $Nu>Nu_{critical}$, some fluctuations will appear in the temperature profiles.

The results from experimental investigation show that the values of both the air temperature and Nusselt number have influence on the output voltage of the CTA. In this study, two ways have been employed to compensate the ambient temperature changes. In the first case, the effect of ambient temperature variation is only considered but in the second case, the effect of Nusselt variation is also regarded. At low temperature variation, the accuracy of two methods is almost the same whereas by increasing the air temperature, the second method which consider the changes in fluid properties, provide more accurate results in compare with the first method.

REFERENCES

1. Benjamin SF, Roberts CA.Measuring flow velocity at elevated temperature with a hot wire anemometer calibrated in cold flowHeat and Mass Transfer 200245703706.

2. Bruun H H. Hot-wire Anemometry, Principles and Signal Analysis. Oxford University Press; 1995.

3. Collis DC, Williams's MJ.Two-dimensional convection from heated wires at low Reynolds numbersJ. Fluid Mech 19596357384.

4. Dehghan Manshaid M, Kazemi Esfeh M. A new approach about heat transfer of hot-wire anemometer. Accepted to publishing in the Applied Mechanics and Materials Journal, 2012.

5. Farhani. F. Ardakani, Experimental study on response of hot wire and cylindrical hot film anemometers operating under varying fluid temperatures. Flow Measurement and Instrumentation 2009.

6. G. Kanevce, S. Oka, Correcting hot-wire readings for influence of fluid temperature variations 1973DISA Info, 15152124.

7. H. Kramers, Heat transfer from spheres to flowing mediaPhysica1946126180.

8. Improved temperature correction in stream Ware. DANTEC DYNAMICS. Technical Note of dynamics, Publication TN04990922002.

9. M. D. Manshadi, 2011The Importance of Turbulence in Assessment of Wind Tunnel Flow Quality", Book chapter 12Wind Tunnels and

Experimental Fluid Dynamics Research, Edited by Jorge Colman Lerner and Ulfilas Boldes, Intech publisher.

10. Öm. H. Lundstr, M. Sandberg, B. Mosfegh, Temperature dependence of convective heat transfer from fine wires in air: A comprehensive experimental investigation with application to temperature compensation in hot-wire anemometry Experimental Thermal and Fluid Science 200732649657.

11. Suminska OA.Application of a constant temperature anemometer for balloon-borne stratospheric turbulence soundings. MSc thesis. University of Rostock; 2008.

CITATION

Mojtaba Dehghan Manshadi and Mohammad Kazemi Esfeh (2012). Analytical and Experimental Investigation about Heat Transfer of Hot-Wire Anemometry, an Overview of Heat Transfer Phenomena, Dr M. Salim Newaz Kazi (Ed.), ISBN: 978-953-51-0827-6, InTech, DOI: 10.5772/51989.

CHAPTER 8

The New Use of Diffusion Theories For The Design of Heat Setting Process In Fabric Drying

Ralph Wai Lam Ip[1] and Elvis Iok Cheong Wan[1]

[1]Department of Mechanical Engineering, The University of Hong Kong, Pokfulam, Hong Kong SAR

INTRODUCTION

Hot air impingement is one of the most widely used methods for material drying. It is also the most traditional drying approach used in industrial process for various kinds of material, such as wood, paper, food, medicine and construction materials. Many research studies have been carried out to see how it is effectively used to process different types of material, and how it is implemented into the design of heat setting machines, such as spray dryers, conveyor dryers, tunnel dryers, fluidized bed dryers and drum dryers.

To study hot air impingement in textile and clothing industries, modeling of porous type fabric drying process would be the key study area. The heat and mass transfer principles are used as tools to assist with the investigation of the hot air impingement mechanism. The mechanism is usually treated as a mass transfer process of the moisture content from the porous material to the impinging air. The transfer of moisture content from the fabric material to the hot air stream is due to a heat transfer process under an in-equilibrium condition. The change of water phases is traditionally described by linear heat transfer equations. As a matter of fact, the driving force in the internal structure of porous materials is not a simple direct proportional relationship between energy exchange and the phase change of the interacting substances, i.e. air and water. Therefore non-linear analytical models based on the physical properties of fabrics will be proposed in this study to provide better simulation results. In the models, the parameters for modeling will be empirically determined and used to describe the drying phenomenon down to microscopic levels. The descriptions will involve the physical and mechanical properties of the drying materials such as mass density, flow viscosity, thermal conductivity, diffusion properties, cohesive properties and flow kinetics. Ip and Wan (2011) have suggested the strategies of using analytical techniques to determine the modeling parameters, and these methods will be investigated in greater depth in this research.

THREE PERIODS OF A FABRIC DRYING CYCLE

Fabric is usually dried up for the purposes of storing or setting. Using thermal energy to dry up and perform setting has been the most traditional and effective method. In this study, heated air is used as a processing agent. Its physical properties will be changed in the gaining of moisture and the loss of thermal energy. The moisture in the fabric will change to vapor after gaining energy from air to create a mass transfer process. The reduction of moisture and increase of fabric temperature is a complicated heat/mass transfer process. Merely using linear conductive and convective heat transfer equations to model the process seems to be inadequate. Diffusion theories are therefore suggested to present the details of the drying process.

"Preheating", "Constant drying" and "Falling drying" are the three periods of a fabric drying cycle as shown in Fig. 1. In the preheating period, most thermal energy is absorbed by water on the fabric surface because air is a poor thermal conductor. The mass transfer rate of water is not high in this period. When more thermal energy is absorbed, water on the fabric surface will change to vapor by evaporation at a rapid mass transfer rate. The water loss rate will keep constant depending upon the air temperature, velocity and atmospheric pressure. As the mass transfer rate of water is constant in this period, it is labeled as constant drying period. While the moisture content in the fabric is going down from the initial i to critical moisture content k, the moisture on the fabric surface starts to separate and form dry/wet regions. Diffusion will appear at the dry/wet regions to form the falling drying period. Diffusion is a slow mass transfer process in comparison with evaporation happened at the second period, and the water transfer rate is correspondingly decreased to form a non-linear drying result until reaching the final moisture content at o.

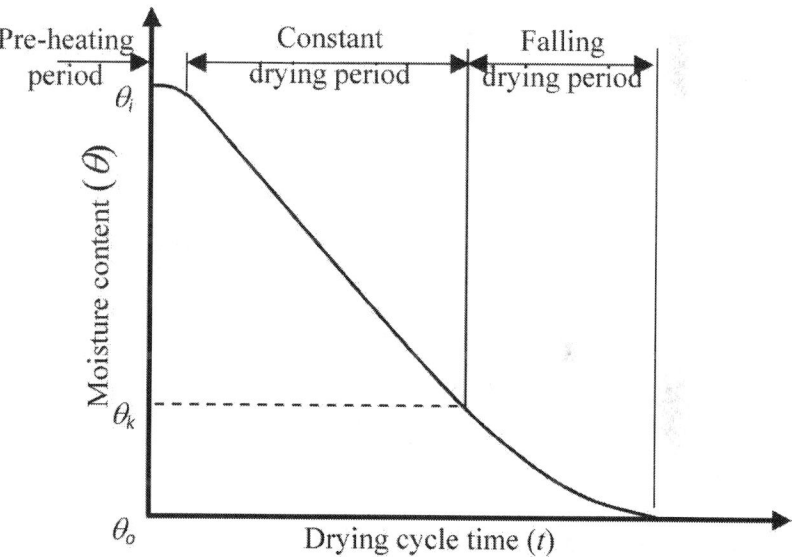

Figure 1: A typical fabric drying curve shows the three drying periods.

Kowalski et al. (2007) has employed partial differentiation and numerical analysis tools to present thermo-mechanical properties of a drying process for porous materials. However, the modeling process is computational intensive and time consuming. It seems impractical to be used in the industrial drying process because a quick response is always needed to manage numerous varying conditions. The concept of using analytical approaches to model drying processes for fruits and sugar has been proved (Khazaei et al., 2008). However, its performance heavily relies upon the sample data and the reliability of the assumptions. The scopes of the research are therefore having rooms to improve the inadequacy. Objectives of the research are to explore and develop robust analytical models that can effectively simulate the characteristics of hot air impingement process for porous fabrics of different textile properties.

Research Objectives

The research objectives in the Chapter are:

- to investigate how to present the drying characteristics of porous type fabric using non-linear analytical models,
- to evaluate the performance of the models in the simulation of drying process, and
- to comment their accuracies in the modeling of different fabric types under various air setting conditions.

In this research, four non-linear analytical models will be studied. The modeling parameters for the models will be empirically determined. The performance of the model will be examined through a careful comparison of testing results. A drying test will be set up to assist the determination of the modeling parameters, and evaluate the performance of the developed models.

Equations for Moisture Flow through Control Volume

Using the approach of control volume to the describe characteristics of moisture flow in porous materials would be close to the phenomenon of fabric drying (Kowalski, 2003). The total mass of the constituents within the control volume will remain unchanged when the porous fabric volume shrinks or otherwise. A set of mass balance equations for each individual constituent, i.e. water, vapor and air in a drying process is given as:

$$\rho^s \, \dot{X}^l = -div \, W^l + {}^l$$

$$\rho^s \, \dot{X}^v = -div \, W^v + {}^v$$

$$\rho^s \, \dot{X}^a = -div \, W^a$$

where ρ is partial mass density, $\dot{X}$ is mass content change rate, is phase transition rate, s is solid fabric, l is liquid phase of water, v is vapor phase of water and a is air. The mass balance Equations (1) – (3) provides a platform for further investigation of moisture content change in a fabric drying process.

Boundary Conditions for the Period of Constant Drying

The water evaporated inside fabric material is much less than that on the boundary surface in the constant drying period. Assuming that the mass flux W of water vapor and air are negligible in this period, the relationship can be given as:

$$W^v = W^a = 0$$

Then, the mass balance Equations (1) – (3) can be rewritten for the calculation of moisture content in the period of constant drying and give:

$$\rho^s \dot{X}^l = -div \; W^l + ^l$$

$$\rho^s \dot{X}^v = ^v << \rho^s \dot{X}^l$$

$$\rho^s \dot{X}^a = 0$$

Equation (5) shows the phase transition of water inside the fabric. Equations (6) and (7) show the phase transition of vapor and air respectively inside the fabric.

Boundary Conditions for the Period of Falling Drying

When the fabric moisture content falls to the critical point at k, meniscoidal water droplets will recede and the drying rate is slowing down until completely dry. The characteristics of fabric drying at this period could be divided into the hygroscopic and non-hygroscopic states, thus, the heat/mass transfer in this period is getting complex. At the initial stage of falling drying, the fabric is fully saturated and water flows in the form of liquid fluxes mainly due to capillary action. Air pockets gradually form at the second stage to replace some of the moisture to form small air bubbles inside the fabric pores. With further drying, the moisture decreases and the size of air bubbles considerably increases that could reduce the rate of heat transfer. As a result, the heat/mass transfer rate is correspondingly reduced since the thermal resistance of air is much higher than water. The drying cycle stops at the tertiary stage when moisture in the hygroscopic regions is totally removed, and a uniform non-hygroscopic property fabric is formed. The mass balance Equations (1) – (3) for the falling drying is rewritten to give:

$$\rho^s \, \dot{X}^v = -div \, W^v +$$

$$\rho^s \, \dot{X}^a = -div \, W^a$$

where in Equations (8) and (9) are all the constituents in the fabric.

Calculation of Mass Fluxes

The mass flux of water in Equation (5) at the constant drying period is given as:

$$W^l = -\Lambda^l \left[\left(\frac{\partial \mu^l}{\partial T^l} \right) T^l + \left(\frac{\partial \mu^l}{\partial \theta^l} \right) \theta^l + \left(\frac{\partial \mu^l}{\partial X^l} \right) X^l - g \right] \tag{1}$$

$$C^l = \left(\frac{\partial \mu^l}{\partial \theta^l} \right) \tag{2}$$

In Equation (11), is the coefficient of water diffusivity, is chemical potential of water, is relative moisture content, T is absolute temperature and g is gravitational acceleration. The equation shows the relationship between water mass flux W^l and the gradients of temperature, volume fraction and mass fraction. The water movement in the fabric is largely due to capillary forces and gravitational force at the constant drying period. C^l is the moisture coefficient related to the moisture cohesive force in the fabric. The mass flux of vapor in Equation (9) for the falling drying period is given as:

$$W^v = -\wedge^v \left[\left(\frac{\partial \mu^v}{\partial T^v} \right) T^v + \sum \left(\frac{\partial \mu^v}{\partial \theta^v} \right) \theta^v \right]$$

(3)

$$C^v = \left(\frac{\partial \mu^v}{\partial \theta^v} \right)$$

(4)

The generation of moisture is due to phase transition of water into vapor, in which, the efflux of vapor is significant. The coefficient C^v for water vapor could be experimentally determined.

FABRIC DRYING TESTS

A series of experiments were conducted to measure the drying characteristics of a group of fabric samples. Cotton is the major studying material in the tests as it is used most widely in clothing industry. The objectives of the experimental tests are to examine the drying characteristics of the fabrics under different boundary conditions, such as fabric texture, density, thickness, air temperature and impingement velocity.

Drying Test Set Up

Six cotton fabric samples were examined. The samples were labeled from A to F, and their properties are listed in Table 1.

The set up as shown in Fig. 2 is an air heater providing hot air stream for each drying test. The temperature and speed of the impinging air are adjustable to provide different boundary conditions for the study. Disc-shaped fabric samples of 100 cm² in area are mounted on a polystyrene backing plate with wire gauze facing the impinging hot air.

Table 1: The properties of the fabric samples for drying tests.

Fabric sample	Fabric texture	Yarn structure	Density (g/m³)	Thickness (mm)
A	Plain knitted	20 s/2	224.4	0.6594
B	Plain knitted	32 s/1	147.7	0.4363
C	Plain knitted	20 s/2	271	0.7769
D	Plain weaved	-	182	0.5638
E	Plain knitted	20 s/1	193	0.5025
F	Plain knitted	32 s/2	200	0.6188

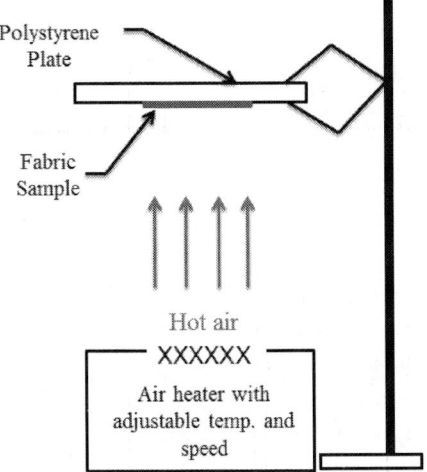

Figure 2: Schematic diagram of the drying test setup.

In the tests, each fabric sample was being dried under eight conditions as listed in Table 2. The fabric weight was measured by an electronic balance every 30 seconds of drying under different setting conditions. The repeated drying and weight measurement procedures were conducted until all the moisture in the fabric samples was removed. The same testing procedures were repeated for all the fabric samples as given in Table 1.

Table 2: The air setting conditions of drying tests for the six fabric samples.

Setting condition	Air temperature (°C)	Impinging velocity (m/s)
1	80	1.48
2	81.5	1.45
3	86.5	1.43
4	54	1.1
5	55.5	1.15
6	54	1.02
7	57	1.41
8	58	1.46

Results and Discussions of the Drying Tests

Fig. 3 shows testing results from the six fabric samples under air setting condition 1 as listed in Table 2. The normalized water contents instead of the absolute values recorded from the tests were used in order to compensate the variation of fabric weight among the six samples.

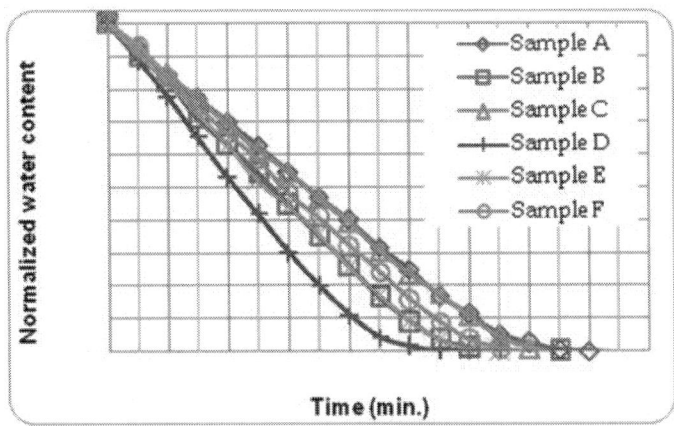

Figure 3: Drying curves for the six fabric samples under air setting condition 1.

The testing results as illustrated in Fig. 3 for the tested fabrics have shown the relationships among water content, fabric density, texture, air temperature and impingement velocity. Their relationships are discussed in the following sections.

Drying Rate versus Fabric Texture

Table 3 lists the results of four fabric samples tested under the air setting conditions of 1, 4, 6 and 8 as listed in Table 2. Among the tested samples, sample D is weaved fabric and the others are knitted fabrics. The drying rate of sample D is the highest among the others at the constant drying period.

Table 3: Drying rate of the four tested samples with different fabric texture.

Air setting condition	**Sample B** - Plain knitted		**Sample D** - Plain weaved	
	1/Drying time (min^{-1})	**Drying rate at constant drying period (g/min)**	**1/Drying time (min^{-1})**	**Drying rate at constant drying period (g/min)**
1	0.083	0.439	0.1	0.462
4	0.08	0.383	0.1	0.395
6	0.095	0.392	0.083	0.398
8	0.065	0.281	0.065	0.287
	Average drying rate = 0.374 g/min		Average drying rate = 0.386 g/min	
	Sample E - Plain knitted		**Sample F** - Plain knitted	
1	0.083	0.431	0.077	0.426
4	0.067	0.304	0.057	0.323
6	0.063	0.349	0.065	0.385
8	0.061	0.29	0.056	0.302
	Average drying rate = 0.344 g/min		Average drying rate = 0.359 g/min	

Drying Rate versus Fabric Density and Thickness

Table 4 lists the testing results of two selected fabric samples A and C with same texture, yarn structure and different density and thickness under the air setting conditions of 1, 4, 6 and 8.

Table 4: Drying rates of samples A and C for different air setting conditions.

| Air setting | Sample A - 224.4 g/m³, 0.6594 mm | | Sample C - 271 g/m³, 0.7769 mm | |
	1/Drying time (min⁻¹)	Drying rate at constant drying period (g/min)	1/Drying time (min⁻¹)	Drying rate at constant drying period (g/min)
1	0.071	0.416	0.067	0.443
4	0.063	0.407	0.063	0.391
6	0.065	0.378	0.053	0.396
8	0.047	0.281	0.044	0.289
	Average drying rate = 0.371 g/min		Average drying rate = 0.380 g/min	

The results listed in Table 4 show a similar result at the constant drying for the fabrics with different density and thickness, but same texture and yarn structure.

Drying Rate versus Air Temperature

Table 5 lists the results of the drying rate for all fabric samples tested under air setting conditions 1 and 8 with similar impingement velocity at 1.47 m/s and different temperature.

Table 5: Drying rate of fabrics under different air temperature.

| Temp. (°C) | Sample A | | Sample B | | Sample C | |
	1/time (min⁻¹)	Drying rate (g/min)	1/time (min⁻¹)	Drying rate (g/min)	1/time (min⁻¹)	Drying rate (g/min)
58	0.047	0.281	0.065	0.281	0.044	0.289
80	0.071	0.416	0.083	0.439	0.067	0.443
Temp. (°C)	Sample D		Sample E		Sample F	
	1/time (min⁻¹)	Drying rate (g/min)	1/time (min⁻¹)	Drying rate (g/min)	1/time (min⁻¹)	Drying rate (g/min)
58	0.065	0.287	0.061	0.29	0.056	0.302
80	0.1	0.462	0.083	0.431	0.077	0.426

The drying rate at the constant drying period increases with the rise of air temperature for all fabric samples.

Summary of the Experimental Findings

The period of constant drying as illustrated in Figs. 1 and 3 has constituted a large portion of the drying cycle. The moisture reduction rate at the period could be used as an indicator to show the properties of the fabric, and conditions of the impinging air. The experimental findings in Tables 3 – 5have shown the performance of the drying process against the boundary conditions including fabric texture, density, thickness, air temperature and impinging velocity. It has been observed that the increase of air temperature and velocity will speed up the drying rate. The fabric properties could also affect the drying rate but not as much as the air properties. These findings could be useful in the setting up of analytical models to simulate each period of a fabric drying cycle.

DEVELOPMENT OF NON-LINEAR ANALYTICAL MODELS TO SIMULATE THE DRYING OF POROUS TYPE FABRICS

As mentioned in Section 2, the drying rate of porous type fabrics has a non-linear relationship with time at the falling drying period. Some inaccurate results will be found if the traditional linear heat transfer equations are applied because the heat transfer coefficient changes with the change of the moisture contents at the falling drying period. To ensure an accurate modeling of the drying process, the heat transfer coefficient should be adjustable corresponding to the diffusion properties in the forming of dry/wet regions as mentioned in Section 2.4. A non-linear model is therefore used to describe the process characteristics (Haghi, 2006; Moropoulou, 2005). The knitted and weaved fabrics studied in this research are considered porous type materials because they contain unidirectional pores. The randomly distributed pores give an environment to establish a diffusion process when portions of water dry up. It is clear that the moisture diffusion rate has a close relationship to the size and number of fabric pores. The studied models given in Section 4.6 will address these essential modeling parameters. The other fabric parameters including texture, density and thickness that correlate to the drying rate will also be modeled by different modeling principles, and given in Sections 4.3 – 4.5.

Determination of the Critical Moisture Content

The two periods of a fabric drying cycle as illustrated in Fig. 1 should be modeled separately. Traditional heat transfer equations could be used as modeling tools for the constant drying period, whilst, non-linear modeling equations should be considered when the moisture reduction rate varies with time in the falling drying period. The critical moisture content k at the beginning of the falling drying period will be the separating point between the two periods. The finding of k is given from the plotting of the normalized drying rate versus the moisture content in gram per gram of the dried fabric, see Fig. 4.

The moisture reduction rate m˙n at the nth period of a time interval Δt can be expressed as:

$$\dot{m}_n = \frac{m_n - m_{n+1}}{\Delta t} \text{ for the first time interval,}$$

(5)

$$\dot{m}_n = \frac{(m_{n-1} - m_n) + (m_n - m_{n+1})}{2\Delta t} \text{ for a time interval n > 0}$$

(6)

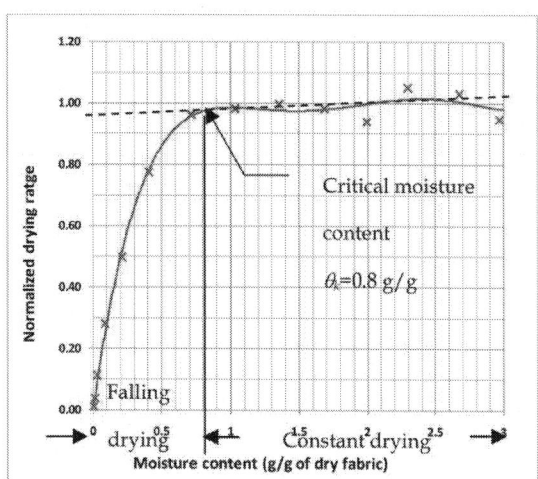

Figure 4: Plotting of normalized drying rate versus moisture content for fabric sample A.

The critical moisture content $_k$ can be identified from the curve as shown in Fig. 4 at the point of dramatic decrease of m'n. The determined $_k$ for fabric sample A under air setting condition 3 listed inTable 2 is 0.8 g/g. The testing results for other fabric samples under the same air setting condition are similar and listed in Table 6.

Table 6: Critical moisture contents of the tested fabric samples.

Fabric sample	Critical moisture content (k (g/g dry fabric weight)
A	0.8
B	0.7
C	0.8
D	0.8
E	0.7
F	0.7

Using Diffusion Theories to Model a Fabric Drying Process

The boundary conditions for heat/mass transfer in the porous fabric have been discussed in Sections 2.2 to 2.5. However, they are not good enough to estimate the fabric moisture content during the drying process. It is necessary to have a further investigation to estimate the moisture content in individual period of drying. The authors have set-up a group of models based upon diffusion theories and Kowalski's (2003) boundary equations to simulate the moisture changing rate under various boundary conditions (Ip and Wan, 2011). The investigated drying models will be presented in differential forms to address the movement of moisture contents in fabric. The models are based on the principles of chemical diffusion mechanism to calculate the rate of moisture change (dM/dt) according to a set of modeling parameters empirically determined from drying cycles (Kowalski, 2000;Schlunder, 2004). Four non-linear analytical models, namely "Kinetics", "Diffusion", "Kinetics model based on the solutions of diffusion equations" and "Wet surface" have been developed, and the principles are given in the following sections.

First Order Kinetics Model

Roberts and Tong (2003) have shown a successful result in the modeling of bread drying process using first order exponential equations. In their research, microwave was used as the drying agent, and the process has been assumed as isothermal. Unfortunately, the experiential setup is quite different from convective drying using impinging air in this study. It is therefore necessary to develop new modeling equations for porous type fabrics. Schlunder (2004) has stressed that the falling drying period should be considered as an isothermal process. First order exponential equations might be appropriate to describe the process, thus, the first model developed in this study is labeled as "First order kinetics model".

In the First order kinetics model, there is an assumption that the vaporization of water inside fabric can be described as a kinetic reaction motion of water molecules. The reaction rate is treated as the moisture reduction rate at the falling drying period. Thus, the water evaporation rate will correlate with the moisture content. The equation of the kinetic model is given as:

$$-\frac{dM}{dt} = kM^n$$

(7)

In Equation (17), M is the instant moisture content and $n = 1$ for the first order kinetics. If M_o is the initial moisture content at the beginning of the falling drying period, i.e. critical moisture content k, the integration result of the differential form Equation (17) will be:

$$\frac{M}{M_o} = e^{-kt}$$

(8)

where k is the kinetic coefficient.

The testing results of moisture content as shown in Fig. 4 are further plotted in terms of drying cycle time t and given in Fig. 5. The red line in the figure represents the drying curve and the black line is the approximated drying rate at the constant drying period.

The kinetic coefficient k in Equation (18) at the falling drying period can be obtained by regenerating a new plotting from the results illustrated in Fig. 5. The ratio of M/Mo in the equation shows an exponential relationship with $-kt$. It can be converted into a linear relationship by applying logarithm for both sides of Equation (18). Fig. 6 shows a plotted graph of $\ln(M/Mo)$ versus the drying cycle time from the experimental records in Fig. 5.

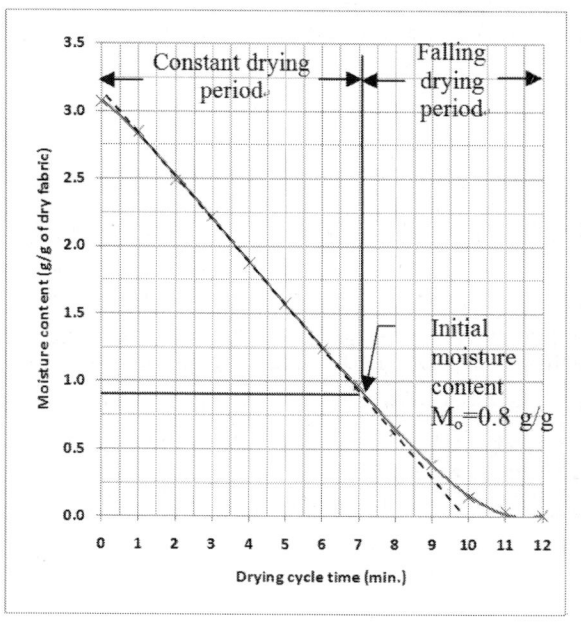

Figure 5: Experimental records of the drying of fabric sample A.

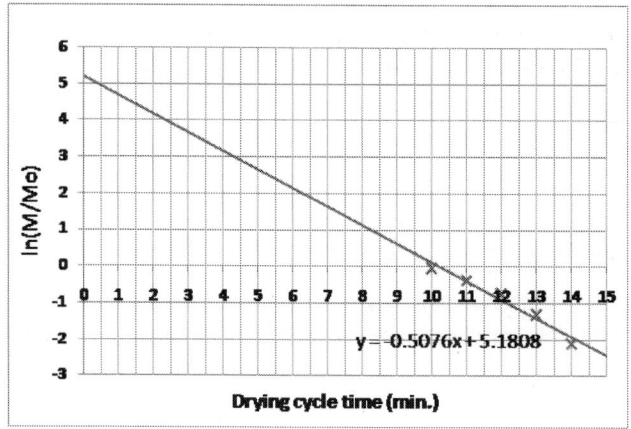

Figure 6: Determination of kinetic coefficient k for fabric sample A.

A graphical method to determine the kinetic coefficient k is to measure the slope of the fitted line inFig. 6. The slope is measured as -0.5076 that will be the tested fabric kinetic coefficient. An alternative method to determine k is to modify Equation (18) using Arrhenius relationship (Roberts and Srikiatden, 2005). It is a common method to correlate rate constant with reaction temperature $T$for kinetics reactions in chemistry. The Arrhenius form of equation in terms of k and T is:

$$k = Ae^{-E_a/RT}$$

(9)

where Ea is the activation energy, R is the universal gas constant at 8.314x10^{-3} kJ/mol K and A is a constant. Equation (19) gives the relationship of kinetic coefficient k in terms of air temperature T only, and does not include the impinging velocity V. However, V is also a key factor in the drying process and its effect could be empirically determined using linear regression methods. The regression equation for the calculation of k from the Arrhenius relationship in Equation (19) is given by taking natural algorithm of the equation.

$$\ln k = \frac{-E_a}{R} \frac{1}{T} + \ln A$$

(10)

A in the Arrhenius equation means reaction per time and is correlated to the impinging velocity in the studied drying models. Thus, the First order kinetics model in Arrhenius form can be written in terms of T and V to give:

$$\ln k = a + b\frac{1}{T} + c\ln V$$

(11)

If the Arrhenius relationship is applied to describe a fabric drying process, a plotting of ln k versus $1/T$will give a straight line. The slope and intercept of the line are used to determine the correlation constants of Ea and b as given in Equations (20) and (21). The kinetic coefficient k of the fabric sample A calculated from Equation (19) under the air setting conditions listed in Table 2 are given inTable 7, and the corresponding values of ln k, $1/T$ and ln V determined from Equation (21) are listed in Table 8.

Table 7: The calculated kinetic coefficients from experiments and Arrhenius equation for fabric sample A (k^* from experiential results, k from Arrhenius equation).

Setting condition	Air temperature (K)	Impinging velocity (m/s)	k^*	k
1	353	1.48	0.5295	0.5198
2	354.5	1.45	0.5818	0.523
3	359.5	1.43	0.5076	0.5336
4	327	1.1	0.5297	0.4634
5	328.5	1.15	0.5494	0.4667
6	327	1.02	0.5537	0.4634
7	330	1.41	0.3589	0.4699
8	331	1.46	0.3568	0.4722

Table 8: The calculated correlation constants for the tested fabric sample A.

Setting condition	ln k	1/T (K⁻¹)	ln V (m/s)
1	-0.6358	0.0028	0.392
2	-0.5416	0.0028	0.3716
3	-0.6781	0.0028	0.3577
4	-0.6354	0.0031	0.0953
5	-0.5989	0.003	0.1398
6	-0.5911	0.0031	0.0198
7	-1.0247	0.003	0.3436
8	-1.0306	0.003	0.3784

Fig. 7 illustrates the plotting of $\ln k$ versus $1/T$ from data in Table 8. Results from the plotting were used to calculate the activation energy Ea and A as given in Equation (20).

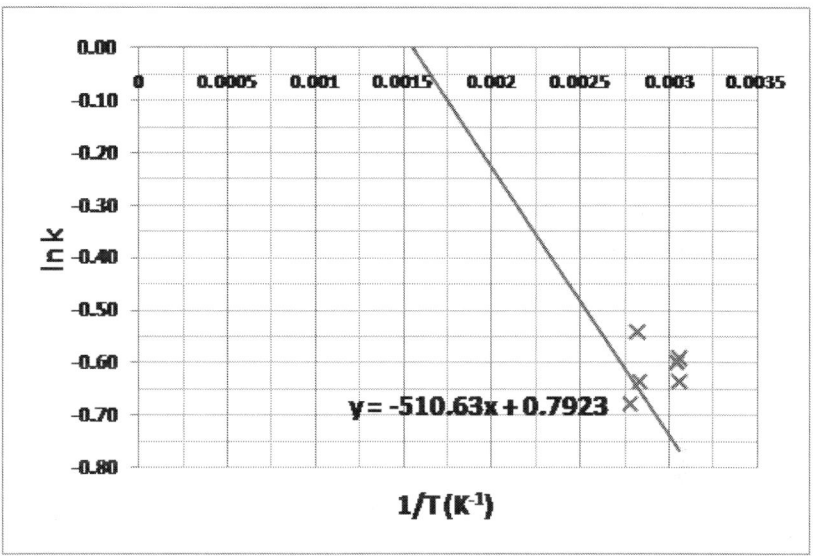

Figure 7: The plotting of ln k versus $1/T$ for fabric A under various air setting conditions.

The calculated values for Ea and A are 4.2454 kJ/mole and 2.2085 respectively given from Fig. 7. The final form of the Arrhenius equation for fabric sample A will be given as:

$$\ln k = -510.63 \frac{1}{T} + 0.7923 \tag{12}$$

Experimental results listed in Table 8 can be further used to determine the coefficients of a, b and c as given in Equation (21) by linear regression methods (Cohen, 2003). The regression results produced from MicroSoft Excel for fabric sample A are given in Fig. 8.

SUMMARY OUTPUT				
Regression Statistics				
Multiple R	0.8912112			
R Square	0.7942575			
Adjusted R Square	0.7256766			
Standard Error	0.1058416			
Observations	9			
ANOVA				
	df	*SS*	*MS*	*F*
Regression	2	0.259478294	0.129739	11.58133
Residual	6	0.067214622	0.011202	
Total	8	0.326692916		
	Coefficients	*Standard Error*	*t Stat*	*P-value*
Intercept	5.5354099	1.333590724	4.150756	0.006006
1/T[k]	-1974.7449	429.2456499	-4.6005	0.00369
lnV	-1.5207901	0.359403041	-4.23143	0.005491

Figure 8: Regression table for fabric sample A determined from the First order kinetics model.

Using results from the regression table, the model equation is given as:

$$\ln k = 5.535 - 1974.7\left(\frac{1}{T}\right) - 1.521 \ln V, \text{where} a = 5.535, b = -1974.7 \text{ and} c = -1.521$$

(13)

Table 9: Regression results of the tested fabric samples.

Fabric sample	a	b	c
A	5.535	-1974.7	-1.521
B	5.74	-1993.9	-1.21
C	3.714	-1476.9	-0.911
D	4.43	-1584.6	-1.276
E	5.58	-2054.5	-0.494
F	6.087	-2223.7	-0.28

Table 9 lists the regression results of all the fabric samples. A comparison of the differences of k determined from Arrhenius equation and regression model for fabric sample A is listed in Table 10.

Table 10: Comparison of kinetic coefficient determined from Arrhenius equation and regression model for fabric sample A.

Setting	Air temp. (°C)	Velocity (m/s)	k1	k2	k1 Dev. (%)	k2 Dev. (%)
1	80	1.48	0.5198	0.5195	1.83	1.9
2	81.5	1.45	0.523	0.5487	10.1	5.68
3	86.5	1.43	0.5336	0.6056	5.12	19.3
4	54	1.1	0.4634	0.5227	12.52	1.32
5	55.5	1.15	0.4667	0.5022	15.06	8.59
6	54	1.02	0.4634	0.5863	16.32	5.89
7	57	1.41	0.4699	0.3786	30.95	5.5
8	58	1.46	0.4722	0.3656	32.34	2.48
		Average	0.489	0.5037	15.53	6.333

The deviations of k for Arrhenius and regression models listed in Table 10 are calculated based on the results obtained from experiments listed in Table 7. The deviations calculated from regression model are much less than Arrhenius equation. The impinging velocity may have been considered in the regression model. Further study about the discrepancy of the modeling results between the regression model and the experimental records was performed.

The red curve illustrated in Fig. 9 is the modeled drying cycle obtained from the regression model for fabric sample A in the falling drying period with a kinetic coefficient k at 0.5037. Discrepancies have been found in comparison with the blue drying curve obtained from experiments. The average discrepancy and standard deviation of the comparison are 14.0139% and 7.8028% respectively.

In conclusion, the falling drying period in a fabric drying cycle can be modeled by the First order kinetics model using an exponential function. A coefficient k for the exponential function can be experimentally determined to model the drying characteristics under various air setting conditions. The coefficient can also be numerically determined from the regression model using air temperature and impinging velocity as the boundary conditions. It has a further relationship to the fabric density and thickness as illustrated in Fig. 10 other than temperature and velocity. Table 11 lists k determined from the regression model for all the fabric samples. It has been found that k decreases with the increase of fabric density and thickness. This relationship could be useful in the estimation of the drying cycle time for the fabrics with different thickness and density.

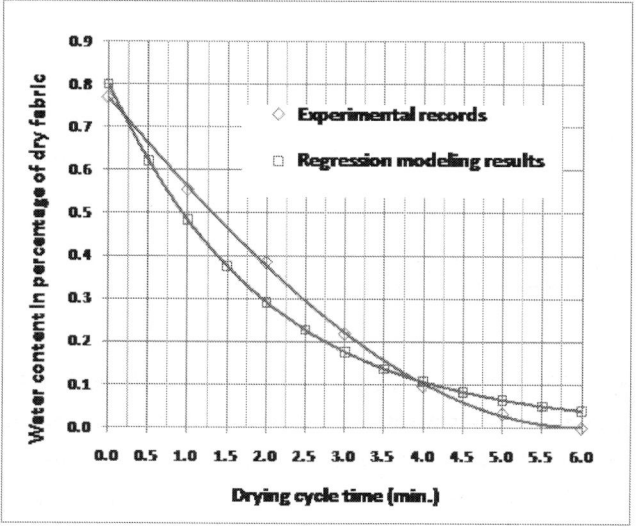

Figure 9: Comparison of experimental records and regression modeling results for fabric sample A.

Table 11: The fabric kinetic coefficient k determined from the regression model.

Fabric sample	Density (g/m³)	Thickness (mm)	k
A	224	0.6594	0.504
B	148	0.4363	0.633
C	271	0.7769	0.414
D	182	0.5638	0.561
E	193	0.5025	0.547
F	200	0.6188	0.587

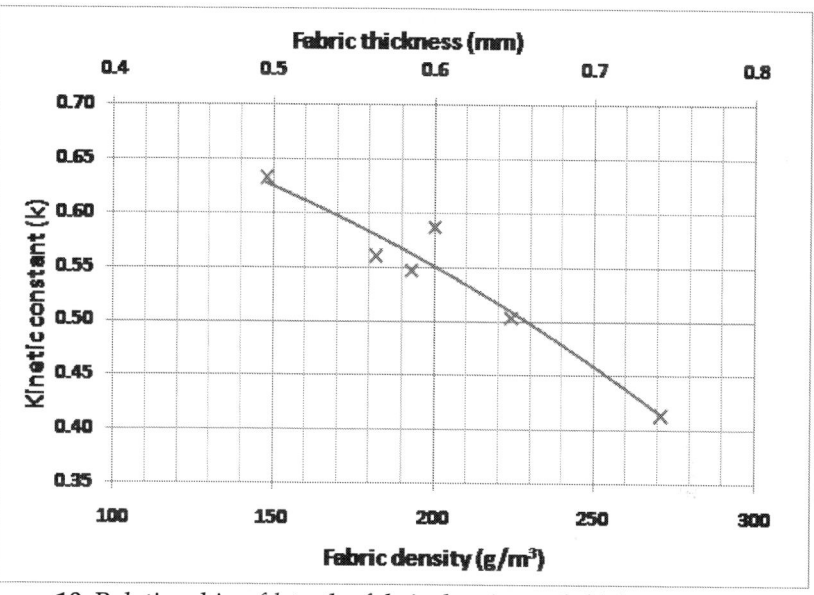

Figure 10: Relationship of k to the fabric density and thickness.

Diffusion Model

Most of the Diffusion models presenting the change of moisture content have been based upon Fick's law (Ramaswamy and Nieuwenhuijzen, 2002). The Fick's first law states the diffusion flux flowing from the regions of higher concentration to lower concentration obeying a magnitude proportional relationship to the concentration gradient. The one dimensional Fick's first law in differential form is given as:

$$J = -D\frac{\partial \phi}{\partial x}$$

(14)

where J is the diffusion flux in $m^{-2}s^{-1}$, D is the effective diffusion coefficient in $m^2\ s^{-1}$, ϕ is the concentration in m^{-3} and x is a linear distance in m. The Fick's second law given in Equation (25) shows the rate of concentration change. It is given by the derivative of Equation (24), with the assumption of D as a constant. The Fick's second law has been used commonly to simulate the drying process of agricultural products (Khazaei et al., 2008), such as seeds and grains.

$$\frac{\partial \phi}{\partial t} = D\frac{\partial^2 \phi}{\partial x^2}$$

(15)

If the Fick's second law is applied to model the process of drying porous fabric, the fabric will be considered as an infinite thin slab and dried from one direction. If heat transfer from the surrounding to the fabric is negligible, the integration result from Equation (25) will give the Diffusion model. The model equation given in Equation (26) is in terms of moisture content M, fabric thickness L, effective diffusion coefficient D and the drying cycle time t.

$$\frac{M}{M_o} = \frac{8}{\pi^2} \exp\left(-\frac{\pi^2 Dt}{4L^2}\right)$$

(16)

The Diffusion equation is similar to Equation (18) of the First order kinetics model. The only difference between the two model equations is the fabric thickness L included in the Diffusion model. As the same as in the First order kinetics model, the effective diffusion coefficient D in Equation (26) can be acquired from the plotting of $\ln (M/Mo)$ versus t as illustrated in Fig. 6. Equation (27) is obtained when a logarithm is applied to both sides of Equation (26).

$$\ln \frac{M}{M_o} = \left(-\frac{\pi^2 D}{4L^2}\right) t - 0.21$$

(17)

The slope of the fitted straight line in Fig. 6 is -0.5076 for fabric sample A with a thickness of 0.6594 mm. D is then calculated by substituting the slope and L back to Equation (27), and the estimated value is 8.945×10^{-8}. The equation form of the Diffusion model and the First order kinetics is similar, D in the Diffusion model can be therefore calculated using the regression model. The Diffusion model in regression form is given in Equation (28).

$$\ln D = a + b\frac{1}{T} + c\ln V$$

(18)

Using information in Table 8 to determine the constants of a, b and c. The determined regression model equation for fabric sample A is given as:

$$\ln D = 20.879 - 12070.75\frac{1}{T} - 2.173\ln V$$

(19)

The regression results for all the fabric samples are listed in Table 12.

Table 12: Regression results of all the fabric samples modeled by Diffusion model.

Fabric sample	a	b	c
A	20.88	-12070.75	2.173
B	10.59	-2006.16	1.215
C	11.43	-1495.8	0.972
D	11.374	-1601.97	1.277
E	10.578	-1895.1	0.526
F	10.08	-2201.5	0.242

The effective diffusion coefficient D for fabric sample A calculated from various air setting conditions are listed in Table 13.

Table 13: Effective diffusion coefficient D for fabric sample A under various conditions.

Setting	Air temp. (°C)	Impinging velocity (m/s)	Effective diffusion coeff. D
1	80	1.48	7.03023×10^{-7}
2	81.5	1.45	8.49438×10^{-7}
3	86.5	1.43	1.40575×10^{-6}
4	54	1.1	8.83487×10^{-8}
5	55.5	1.15	9.49414×10^{-8}
6	54	1.02	1.04100×10^{-7}
7	57	1.41	7.20528×10^{-8}
8	58	1.46	7.46037×10^{-8}
		Average	4.24032×10^{-7}

Fig. 11 illustrates a comparison between the experiential records and the modeling results from regression model using D at $4.24032\text{x}10^{-7}$.

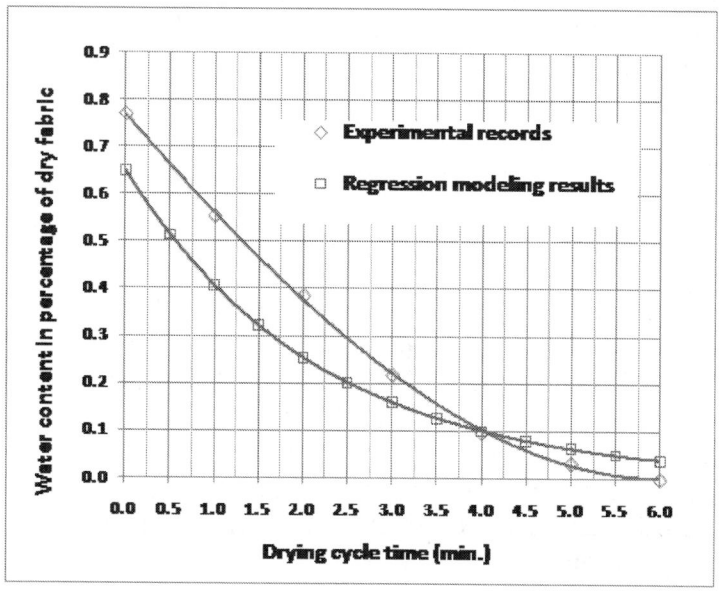

Figure 11: A comparison of regression results from Diffusion model and experimental records for fabric sample A.

The Diffusion model has been applied to model each of the drying cycle for all fabric samples, the discrepancies between the modeling results and records from experiments are listed in Table 14.

Table 14: Discrepancies of a comparison of Diffusion model and records of experiments.

Fabric sample	A	B	C	D	E	F
Discrepancy (%)	76.6	42.9	49	44.4	48.4	43.3

Kinetics Model Based on the Solutions of Diffusion Equations

Fick's second law for diffusion applications is commonly used to simulate mass transfer process in convective drying. However, the exponential term in Equation (26) causes a restriction to the Diffusion model be applied in the falling drying period. A separate modeling process is needed to describe the constant drying period for completed modeling of a drying cycle. Efremov (1998, 2002) has proposed a mathematical solution to solve the Frick's law using integral error functions:

$$\frac{m}{m_o} = 1 - erf\frac{x}{2\sqrt{Dt}} \tag{20}$$

$$\frac{m}{m_o} = 1 - N_o\frac{t}{m_o} \tag{21}$$

where m is the moisture removal rate, mo is the initial moisture removal rate, No is the drying rate at the constant drying period and x is the fabric thickness. Substituting the boundary conditions of $t = 0$ and $t = t_f$ for a drying cycle at the starting and ending points, a kinetics model equation developed from the diffusion model is given as:

$$\frac{m}{m_o} = 1 - N_o\frac{t}{w_o} + \frac{N_o\sigma\sqrt{\pi}}{2w_o}\left(1 - erf\frac{t_f - t}{\sigma}\right) \tag{22}$$

where σ is a characteristic drying time and expressed as:

$$\sigma = 2\frac{N_o t_f - m_o}{N_o\sqrt{\pi}} \tag{23}$$

The first and second terms in the right-hand-side of Equation (32) represent the characteristics in the constant drying period, and the third term represents the falling drying period. The new kinetics equation consists of two modeling sections to describe the linear and non-linear parts of a drying process. The drying rate at constant drying period No and the drying cycle time t_f are needed to be predetermined when Equation (32) is applied. No is the slope of the dotted line as shown in Fig. 5, they are given in Table 15.

Table 15: The drying rate at constant drying period of the fabric samples under different air setting conditions.

Setting condition	Drying rate at constant drying period of the fabric samples (g/s)					
	A	B	C	D	E	F
1	0.0069	0.0073	0.0074	0.0077	0.0072	0.0071
2	0.0069	0.0079	0.0083	0.008	0.0084	0.0071
3	0.0065	0.0067	0.0069	0.0069	0.007	0.0065
4	0.0068	0.0064	0.0065	0.0066	0.0051	0.0054
5	0.0066	0.0065	0.0072	0.0066	0.0074	0.0074
6	0.0063	0.0065	0.0066	0.0066	0.0058	0.0064
7	0.0047	0.0046	0.0051	0.0047	0.0048	0.0051
8	0.0047	0.0047	0.0048	0.0048	0.0048	0.005
Average	0.00618	0.00633	0.0066	0.00649	0.00631	0.00625

The drying rate at constant drying period N_o listed in Table 15 for each fabric sample and their corresponding drying cycle time t_f are employed to assist the simulation of entire drying process. Equation (32) is the modeling tool to calculate the moisture removal rate m from to to t_f. The red curve in Fig. 12 illustrates the modeling results determined from Equation (32). The values for N_o and t_f are 0.00618 and 15 min. respectively. The modeling process has been repeated for the falling drying period using the final term of Equation (32), and the results are given in Fig. 13.

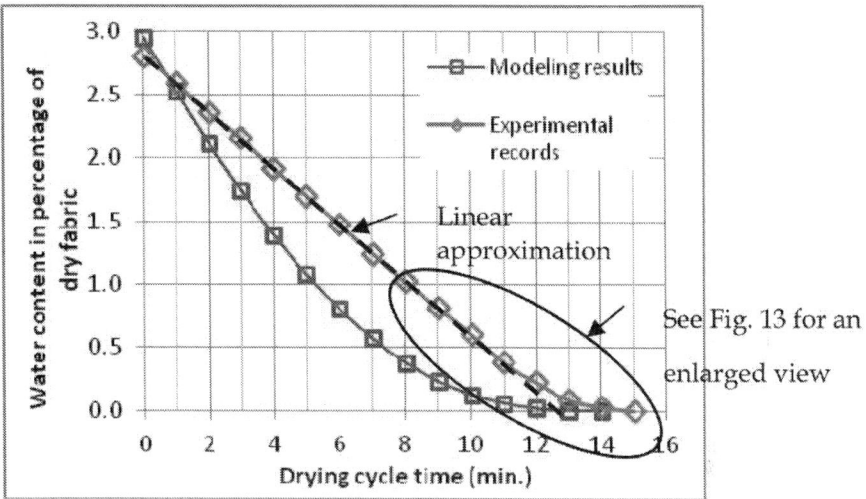

Figure 12: A comparison of modeling results from Equation (32) and testing records of a complete drying cycle for fabric sample A under air setting condition 3.

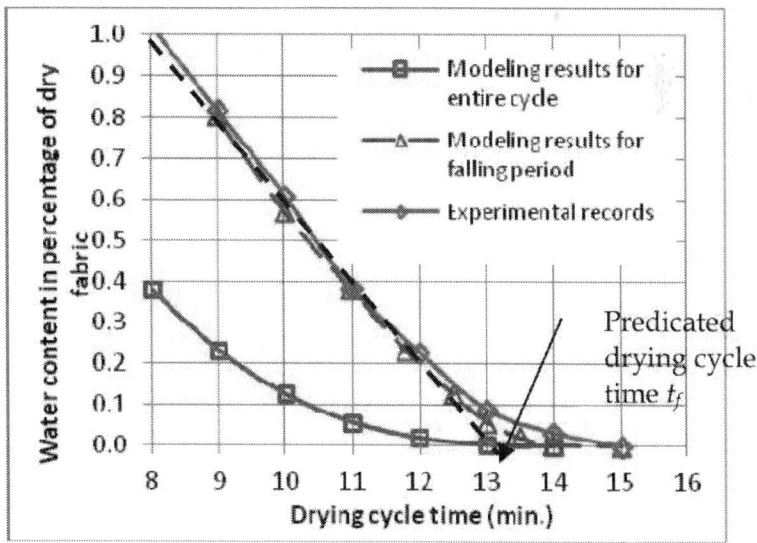

Figure 13: A comparison of modeling results from Equation (32) for the falling drying period and testing records of fabric sample A under air setting condition 3.

Tables 16 and 17 list the discrepancies of the two modeling results from Figs. 12and 13.

Table 16: Discrepancies between the tests and modeling results for entire drying cycle.

Fabric sample	A	B	C	D	E	F
Discrepancy (%)	41.17	30.25	40.53	30.21	36.63	38.06

Table 17: Discrepancies between the tests and modeling results at the falling drying period.

Fabric sample	A	B	C	D	E	F
Discrepancy (%)	8.7	7.92	10.12	11.69	8.87	7.15

Wet Surface Model

The fourth analytical model, Wet surface, was proposed by Schlunder (1988, 2004). He has proved that the remaining moisture in porous materials would be 20 to 30 % of the saturated moisture content at the initial stage when a drying process reaches the critical moisture stage. The material surface is unlikely to be fully wetted at this low moisture content condition. The Wet surface model is therefore designed to address the characteristics of the partially wet surface. The drying rate M'v is modeled directly in contrast with the calculation of the moisture content M required in the previous models. The Wet surface equation is given as:

$$\frac{\dot{M}_v}{\dot{M}_{vI}} = \frac{1}{1+\phi} \tag{24}$$

$$\phi = \frac{2r}{\pi\varepsilon}\sqrt{\frac{\pi}{4\varphi}}\left(\sqrt{\frac{\pi}{4\varphi}} - 1\right) \tag{25}$$

M'_v is the instant drying rate and M'_{vI} is initial drying rate when the fabric surface is fully wetted, r is pore size, is viscous sub-layer thickness and φ is the fraction of a wet surface. It is assumed that the wet surface fraction is proportional to a ratio of moisture content to the critical moisture content c in the falling drying period, thus the Equation (35) is rewritten as:

$$\phi = \frac{2r}{\pi \varepsilon} \sqrt{\frac{\pi}{4\frac{\theta}{\theta_c}}} \left(\sqrt{\frac{\pi}{4\frac{\theta}{\theta_c}}} - 1 \right)$$

(26)

The results of converting Equation (36) from the critical moisture content c to the final moisture content give an expression as:

$$\theta - \theta_c + \frac{r}{2\varepsilon}\theta_c \ln\frac{\theta}{\theta_c} - \frac{2r}{\sqrt{\pi\varepsilon}}\sqrt{\theta_c\theta} + \frac{2r}{\sqrt{\pi\varepsilon}}\theta_c = \dot{M}_{vI}t$$

(27)

In Equation (37), the parameters for fabric drying modeling are the initial drying rate M'_{vI}, critical moisture content c, pore size r and viscous sub-layer thickness. In fact, M'_{vI} and c can be experimentally determined as discussed in Section 4.1. The viscous sub-layer thickness is in terms of thermal conductivity of air c and heat transfer coefficient U, they are given as:

$$\varepsilon = c/U$$

(28)

$$U = \frac{\lambda N_o}{A(T_h - T_{SL})}$$

(29)

where λ is the latent heat of vaporization, A is the fabric surface area, Th and TSL are temperature of the drying air and the temperature at saturated condition respectively. The values of λ, TSL and c can be found from engineering handbooks. Table 18 lists the calculated for the fabric sample A from Equations (38) and (39).

Table 18: The viscous sub-layer thickness for the fabric sample A.

Setting condition	N_o (g/s)	T_h (K)	T_{SL} (K)	(J/g)	U (W/m²K)	C (W/mK)	(mm)
1	0.0069	353	301	2310	30.79	0.028	0.91
2	0.0077	354.5	301	2304	33.29	0.028	0.84
3	0.0065	359.5	302	2291	25.7	0.028	1.09
4	0.0068	327	293	2372	47.35	0.028	0.59
5	0.0066	328.5	294	2372	45.32	0.028	0.62
6	0.0063	327	293	2372	43.89	0.028	0.64
7	0.0047	330	294	2366	31.04	0.028	0.9
8	0.0047	331	295	2366	30.75	0.028	0.91
						Average	0.8125

The final parameter to be determined for Equation (37) is the fabric pore size r. The pore size can be measured under microscope but it is not practical to determine through microscopic images. A better way to acquire the pore size can be done by graphical methods. Fig. 14 shows the plotting of the predicted drying rate from the Wet surface model versus the moisture reduction rate experimentally determined from Equation (15).

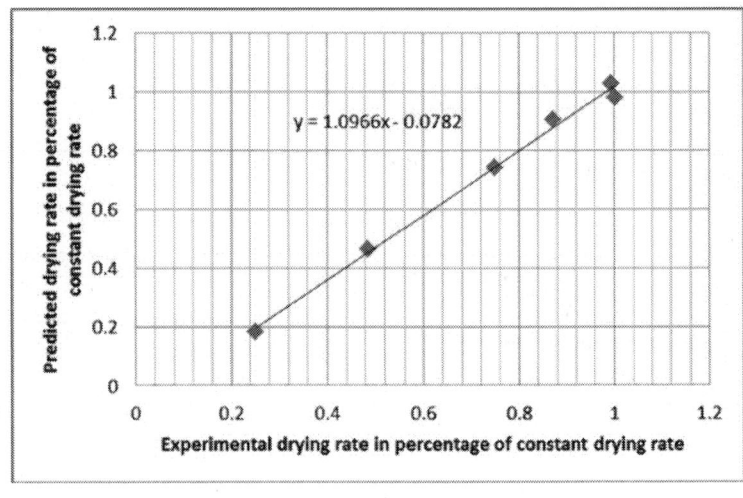

(a)

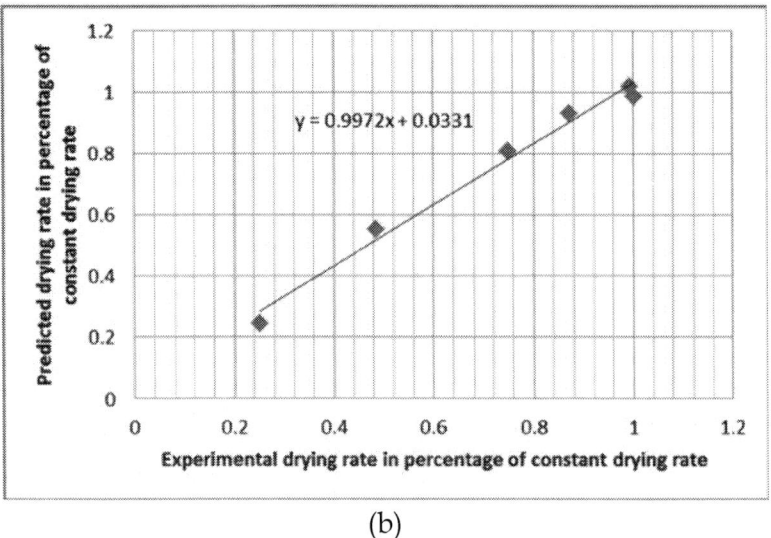

(b)

Figure 14: Relationship of predicted drying rate to experimentally determined drying rate for the assigned pore size r at (a) 0.5 mm and (b) 0.3478 mm for fabric sample A.

The fitted line in Fig. 14(a) shows the results from an assigned pore size of 0.5 mm. Results from the plotting have shown that the slope of the fitted line is not a unity, and the y-intercept does not meet the origin. As a result, the calculated viscous sub-layer thickness from the pore size does not consist with the calculated value as listed in Table 18. Thus, a new assignment of 0.3478 mm was used to create another plotting. The new fitted line as illustrated in Fig. 14(b) is much closer to unity slope and zero y-intercept conditions. The new pore size could be used as the modeling parameter in the Wet surface model.

Table 19 lists the determined pore sizes for other fabric samples. The calculated pore sizes for each fabric sample are then substituted into Equation (37) to determine the drying rate M'_v. A comparison of the discrepancies between the modeled drying rate and recorded data from experimental tests for fabric sample A under the air setting condition 3 is given in Fig. 15.

Table 19: The determined pore size in mm for the tested fabric samples.

| Setting condition | Fabric sample | | | | | |
	A	B	C	D	E	F
1	0.4	0.61	0.19	0.31	0.46	0.36
2	0.4	0.55	0.35	0.7	0.4	0.45
3	0.35	0.4	0.55	0.31	0.21	1.55
4	0.16	0.23	0.22	0.25	0.2	0.5
5	0.19	0.31	0.22	0.4	0.37	0.35
6	0.33	0.07	0.49	0.4	0.37	0.44
7	0.4	1.1	0.6	0.31	0.43	0.55
8	0.55	0.53	0.49	1.5	1.1	1.2
Average	0.3478	0.475	0.3888	0.5225	0.4425	0.675

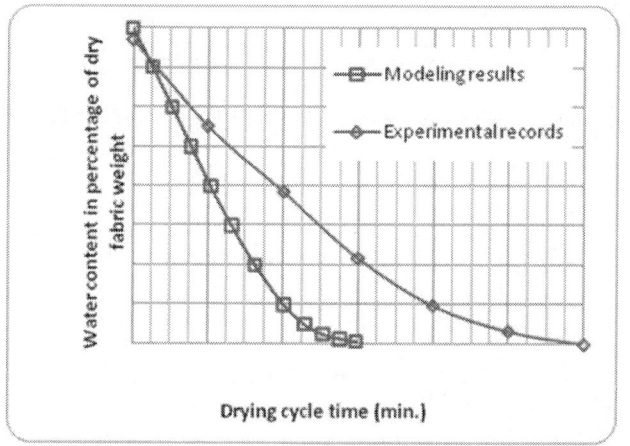

Figure 15: A comparison of the modeling results using Equation (37) and testing records for fabric sample A under test setting 3 for the falling drying period.

The findings given in Fig. 15 have shown that the Wet surface model could not produce an accurate modeling result as the measured discrepancy is 77.53 % based upon the experimental records.

A PERFORMANCE EVALUATION OF THE STUDIED MODELS

The performance of the studied analytical models for the modeling of porous fabric drying process should be reviewed. The percentage of discrepancies from each modeling results in comparison with the experimental records are summarized in Table 20. It is made clear that the Kinetics model based on the solutions of diffusion equations has produced the best performance. The First order kinetics model has provided a better performance than the Diffusion model, and the Wet surface model has given the largest discrepancy from the statistical records listed in Table 20.

Table 20: Summary of performance evaluation of the studied models.

Model	Fabric sample						Average
	A	B	C	D	E	F	
First order kinetics	42.39	34.69	38.74	34.21	42.25	37.11	38.23
Diffusion	76.63	42.86	49.03	44.44	48.35	43.34	50.77
Kinetics based on the solutions of diffusion equations	8.7	7.92	10.12	11.69	8.87	7.15	9.08
Wet surface	77.53	37.47	93.5	56	69.87	64.02	66.4

The findings have shown that the Kinetics model based on the solutions of diffusion equations could be the best one in the simulation of a porous fabric drying process among the others. The required condition for the model is to have a predictable drying cycle time t_f that could be obtained by a linear approximation of the experimentally determined drying curve as shown in Fig. 13. First order kinetics model has also produced a good performance in comparison with the Diffusion and Wet surface model. An empirically determined kinetic coefficient k is only needed for the modeling process, and the coefficient is highly correlated to temperature and impinging velocity of air. A less accurate modeling result observed from the Diffusion model is that a fabric thickness L is needed for the model equation. A significant change of the exponential index in Equation (26) would cause a large discrepancy in the modeling result if the thickness is inaccurately defined. It is understandable that the complexity of finding the pore size would cause unavoidable errors in the Wet surface model. To design experimental strategies determining the pore size to improve the model performance is some further work.

CONCLUSION

The principles of water mass movement due to phase change in the drying of porous fabrics have been studied. The boundary equations for mass transfer between water, vapor and air were used to support the establishing a new set of drying models using diffusion theories. Experiments were done to find information for the determination of the modeling parameters. The performance of the developed models has been evaluated. Among the four models, the Kinetics model based on the solutions of diffusion equations has produced the best performance. In the real life applications, they could act as a mathematical tool to assist a precise estimation of the moisture content in fabric drying or heat setting process under various processing conditions. Further work has been started to apply the developed drying models in the design of garment setting machines for clothing industry.

REFERENCES

1. A. K. Haghi, 2006 Transport Phenomena in Porous Media: A Review. Theoretical Foundations of Chemical Engineering. 40 1 1426
2. A. Moropoulou, 2005 Drying Kinetics of Some Building Materials. Brazilian Journal of Chemical Engineering, 22 2 203208
3. E. U. Schlunder, 1988 On the Mechanism of the Constant Drying Rate Period and Its Relevance to Diffusion Controlled Catalytic Gas-Phase Reactions. Chemical Engineering Science, 43 10 26852688
4. E. U. Schlunder, 2004 Drying of Porous Material during the Constant and the Falling Rate Period: A Critical Review of Existing Hypotheses. Drying Technology, 22 6 15171532

5. G. I. Efremov, 1998 Kinetics of Convective Drying of Fibre Materials Based on Solution of a Diffusion Equation. Fibre Chemistry, 30 6 417422

6. G. I. Efremov, 2002 Drying Kinetics Derived from Diffusion Equation with Flux-Type Boundary Conditions. Drying Technology, 20 1 5566

7. H. S. Ramaswamy, N. H. van Nieuwenhuijzen, 2002 Evaluation and Modeling of Two-Stage Osmo-Convective Drying of Apple Slices. Drying Technology, 20 3 651667

8. J. Cohen, P. Cohen, S. G. West, L. S. Aiken, 2003 Applied Multiple Regression/Correlation Analysis for the Behavioral Sciences. (2nd ed.): Lawrence Erlbaum Associates, Hillsdale, NJ

9. J. Khazaei, G. R. Chegini, M. Bakhshiani, 2008 A Novel Alternative Method for Modeling the Effects of Air Temperature and Slice Thickness on Quality and Drying Kinetics of Tomato Slices: Superposition Technique. Drying Technology, 26 6 759775

10. J. S. Roberts, C. H. Tong, 2003 Drying Kinetics of Hygroscopic Porous Materials under Isothermal Conditions and the Use of a First-Order Reaction Kinetic Model for Predicting Drying. International Journal of Food Properties. 6 3 355367.

11. J. S. Roberts, J. Srikiatden, 2005 Moisture Loss Kinetics of Apple During Convective Hot Air and Isothermal Drying. International Journal of Food Properties, 8 3 493512

12. R. W. L. Ip, I. C. Wan, 2011 New Use Heat Transfer Theories for the Design of Heat Setting Machines for Precise Post-Treatment of Dyed Fabrics. Journal Defect and Diffusion Forum, Vols. 312-315, 748751

13. S. J. Kowalski, 2000 Toward a Thermodynamics and Mechanics of Drying Processes. Chemical Engineering Science, 55 7 12891304 .

14. S. J. Kowalski, 2003 Thermomechanics of Drying Processes, Springer, 3-54000-412-2 Germany

15. S. J. Kowalski, G. Musielak, J. Banaszak, 2007 Experimental Validation of the Heat and Mass Transfer Model for Convective Drying. Drying Technology, 25 1-3 , 107121

CITATION

Ralph Wai Lam Ip and Elvis Iok Cheong Wan (2012). The New Use of Diffusion Theories for the Design of Heat Setting Process in Fabric Drying, Advances in Modeling of Fluid Dynamics, Dr. Chaoqun Liu (Ed.), ISBN: 978-953-51-0834-4, InTech, DOI: 10.5772/48484.

CHAPTER 9

Evaluation of Characteristics of Phase Change Heat Transfer In Ultrafine Cryoprobe

Junnosuke Okajima, Atsuki Komiya, Shigenao Maruyama

Institute of Fluid Science, Tohoku University, Sendai, Japan

ABSTRACT

To reduce the invasiveness of cryosurgery, a miniaturized cryoprobe is necessary. The authors have developed an ultrafine cryoprobe for realizing low-invasive cryosurgery by local freezing. The objectives of this study are to estimate the heat transfer coefficient and investigate the characteristics of the phase change heat transfer in the ultrafine cryoprobe. This cryoprobe has a double-tube structure consisting of two stainless steel microtubes. The outer diameter of the cryoprobe was 550 μm. The alternative Freon HFC-23, which has a boiling point of −82°C at 0.1 MPa, was used as a refrigerant. To evaluate the characteristics of boiling flow in the cryoprobe, the heat transfer coefficient was estimated. The derived heat transfer coefficient was higher than that obtained from the conventional correlation. Additionally, a bubble expansion model was introduced to evaluate the heat transfer mode of the phase change flow in the ultrafine cryoprobe. This model can estimate the liquid film thickness during the expansion of a single bubble in a microchannel. The experimentally measured wall superheat was much lower than that obtained from the model. Therefore, this result also implied that the heat transfer mode in the ultrafine cryoprobe should be nucleate boiling.

Keywords
Phase Change Heat Transfer, Biomedical Application, Cryosurgery, Microchannel

INTRODUCTION

Cryosurgery is a surgical treatment that uses the characteristics of frozen biological tissue to remove undesirable tissues using a cooling device called a cryoprobe. Cryosurgery is less invasive and offers the advantages of low bleeding and a short recovery period [1] .

Some cryoprobes have succeeded commercially, e.g., Accuprobe from Cryomedical Science [2] , CRYOcare from ENDOcare [3] , ERBOKRYO from ERBE [4] , CryoHit from Galil Medical [5] , and SurgiFrost from CryoCath Technologies [6] . Conventional cryoprobes, which are classified into two types depending on the cooling method used, are 3 - 8 mm in diameter. Hence, it has been difficult to treat small lesions, such as pigments on the skin, wrinkles around the eyes, and early breast cancer, by conventional cryoprobes.

Several cryoprobes have been proposed that reduce the level of invasiveness. Takeda et al. [7] developed a Peltier cryoprobe. By changing the electric current supplied to a Peltier module, the surface temperature of the cooling section can be controlled precisely. Aihara et al. [8] developed a flexible, long, slender cryoprobe with vacuum insulation. Boiling heat transfer of an impinging jet using liquid nitrogen was used as the heat transfer mechanism. Additionally, Maruyama et al. [9] developed a flexible cryoprobe that uses the Peltier effect. This cryoprobe consists of a flexible plastic tube. Additionally, miniaturized cryoprobes of around 1 mm diameter have been studied by Benita and Condé [10] and Zhang et al. [11] ; however, the cooling power of those cryoprobes was insufficient because the heat transfer rate decreased owing to the large ratio of the surface area to the volume. Generally, the miniaturization of channel size causes increasing the heat transfer coefficient and decreasing the heat transfer rate comparing to the normal size. Therefore, a mechanism to achieve the both of high heat transfer coefficient and transporting large amount of heat is required and the solution is using the phase change heat transfer.

To overcome this problem, the authors have studied the phase change heat transfer in a co-axial small double tube [12] and developed an ultrafine cryoprobe with boiling heat transfer in a microchannel [13] [14] . The outer diameter of this cryoprobe is 550 µm. The ultrafine cryoprobe may realize new cryosurgery for small lesions with minimum invasiveness, or in a blood vessel using a catheter. In previous study, the theoretical limitation of freezing size by the ultrafine cryoprobe [14] and the refrigerant state inside the cryoprobe [13] were discussed. To realize small-scale cryosurgery using the ultrafine cryoprobe, its cooling performance needs to be clarified. Especially, the heat transfer coefficient is the important parameter to evaluate the cooling performance. The objectives of this study are to estimate the heat transfer coefficient and evaluate the characteristics of the phase change heat transfer in the ultrafine cryoprobe by using the estimated heat transfer coefficient.

EXPERIMENTAL SYSTEM

Ultrafine Cryoprobe

Figure 1 shows the concept of the ultrafine cryoprobe, which consists of inner and outer tubes. The inner tube has a 150 µm outer diameter (OD) and a 70 µm inner diameter (ID). The outer tube has a 550 µm OD and 300 µm ID. Both tubes are made of stainless steel. HFC-23, which is an alternative Freon with a normal boiling point of −82.1°C, was used as a refrigerant. HFC-23 was transported to the inner tube in the liquid state.

The inner tube acts as a capillary for depressurization. Therefore, a large pressure drop occurs in the inner tube, and the temperature of the refrigerant decreases. Furthermore, the refrigerant expands upon exiting the inner tube. Here, the refrigerant changes to a two-phase flow, and the outer tube is cooled. The advantages of this

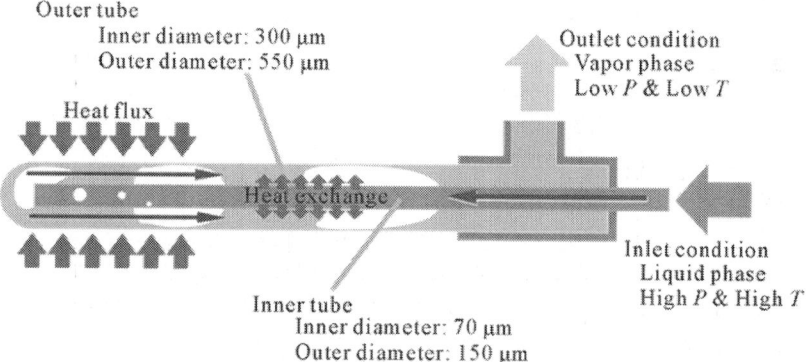

Figure 1: Concept of ultrafine cryoprobe.

cooling system are as follows:

1) The refrigerant can be transported from the reservoir to the cooling section. Therefore, a complex insulation system such as that used in liquid nitrogen cryoprobes is not required.
2) The ultrafine cryoprobe is cooled by boiling heat transfer. Therefore, a higher heat transfer coefficient than that of Joule-Thomson cryoprobes can be expected.
3) Generally, the cryoprobe is required to generate lower than −20°C in order to necrotize an affected area. The size of necrosis area and required heat flux can be predicted by solving the freezing phenomena of biological tissue.

Experimental System

Figure 2 shows the experimental apparatus, which consists of an HFC-23 cylinder, a precooler, several valves, and the cooling section. The measurements were made using T-type thermocouples, a pressure gauge, and a thermal mass flowmeter. The locations of each sensor are shown in Figure 2(b), which also shows the details of the ultrafine cryoprobe section. The refrigerant temperature and pressure at the inlet and outlet were measured by a T-type thermocouple 76 μm in diameter and the pressure transducer, respectively. Nine T-type thermocouples were installed on the ultrafine cryoprobe surface using a thermally conductive, electrically insulating adhesive. Moreover, a leading wire was soldered to the cryoprobe's surface to supply the electrical voltage for controlling the heat flux. The surface of the ultrafine cryoprobe was insulated by Styrofoam [Dow Chemical Company, thermal conductivity = 0.03 W/(m · K)].

Before the experiment, the air in the experimental apparatus was evacuated by a vacuum pump. During evacuation, the needle section was heated by applying a direct current to remove any remaining water vapor. The HFC-23 tank was kept at room temperature. The precooler was cooled by ice water. The HFC-23 vapor from the tank passed through the precooler, and HFC-23 condensed from the vapor phase to the liquid phase. Before the experiment began, Valve 2 was opened to bring the pressure inside the needle section to 0.1 MPa and release the pressure inside the cryoprobe. Then the experiment was started by opening Valve 1.

EXPERIMENTAL RESULTS

Figure 3 shows the time variation in the temperature and mass flow rate without heating by the electric current. The surface temperature of the ultrafine cryoprobe decreased immediately and then remained constant at around −50°C. In previous study [13] , the vapor quality in the tip of cryoprobe was estimated as the slightly subcooled liquid. By assuming that the refrigerant in outer tube is saturated, the pressure in the tip of cryoprobe is estimated at around 0.5 MPa. Therefore, the pressure drop in the inner tube is 3.7 MPa. Additionally, the refrigerant temperature in the outlet decreased gradually. This result indicates that the refrigerant exhaust from the outer tube was capable of cooling.

Figure 4 shows the time variation in the surface temperature with a heat flux of 100 kW/m². The temperature increased by only 2 K after heating by the electric current. Figure 5 shows the time variation in the surface temperature with a heat flux of 170 kW/m².

In this case, the temperature increased by 5 to 7 K after heating by the electric current. Additionally, the temperature response at TC9 was slightly fluctuated. This fluctuation indicated that the partial dryout should occur in the cryoprobe. These two figures proves that the phase change heat transfer of HFC-23 has a large cooling power.

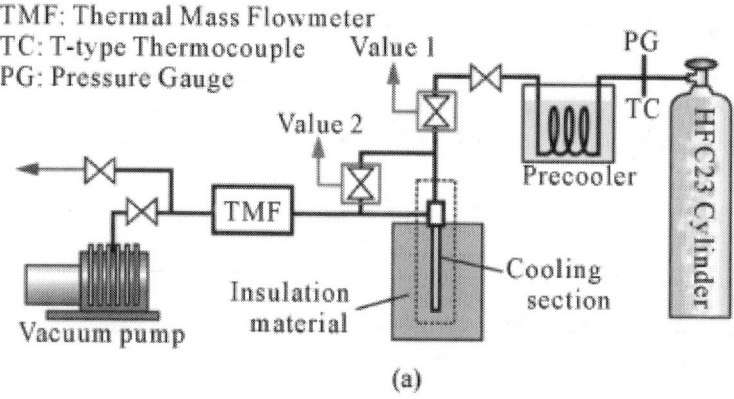

(a)

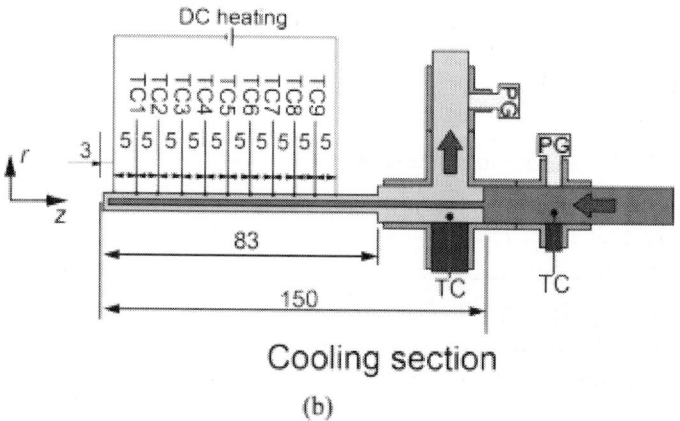

Cooling section

(b)

Figure 2: Schematic of experimental system, (a) Overview of system, (b) Detail of cooling section.

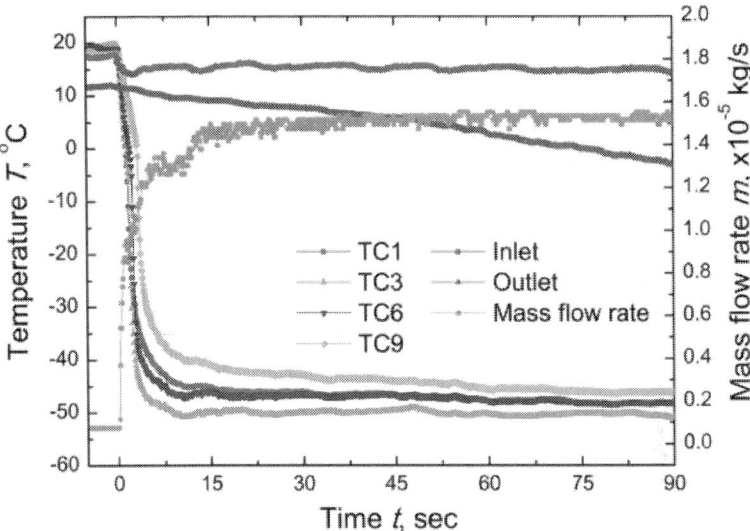

Figure 3: Time variation in temperature at TC1, TC3, TC6, TC9, inlet, and outlet and mass flow rate [14].

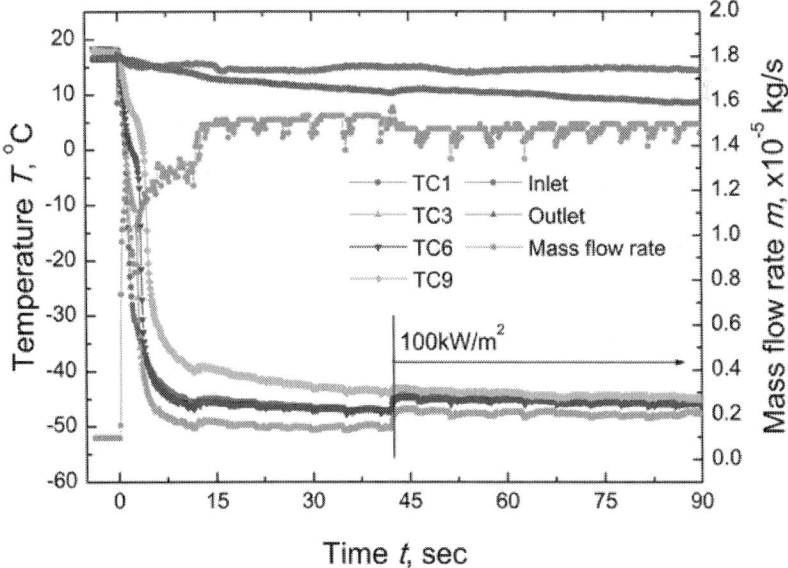

Figure 4: Time variation in temperature at TC1, TC3, TC6, TC9, inlet, and outlet and mass flow rate under heat flux of 100 kW/m² [14].

Figure 6 shows the time variation in the surface temperature with a heat flux of 180 kW/m². The temperature increased by only 5 to 7 K after heating by the electric current. This trend was as same as the case of 170 kW/m². However, as shown in Figure 7, the unstable response of the time variation in the surface temperature was observed at TC8 when a heat flux of 180 kW/m² was supplied. These unstable phenomena occurred accidentally. As shown in Figure 7, the instability started on the downstream side. The temperature at TC9 was around −5°C. In this location, the refrigerant was completely superheated by the external heat flux. On the other hand, a constant temperature was maintained near the tip of the ultrafine cryoprobe. Therefore, the edge of liquid film should exist around the location of TC8 and the wetting and drying should occur periodically at this location.

This temperature fluctuation shown in Figure 7 should not affect to the freezing process in the cryosurgery. The temperature of frozen region should not follow the fluctuation because the heat capacity of frozen tissue is large. The problem in the real application is the refrigerant temperature and time-averaged heat transfer coefficient. Next, the heat transfer coefficient of the ultrafine cryoprobe was derived using the stable temperature data under heating.

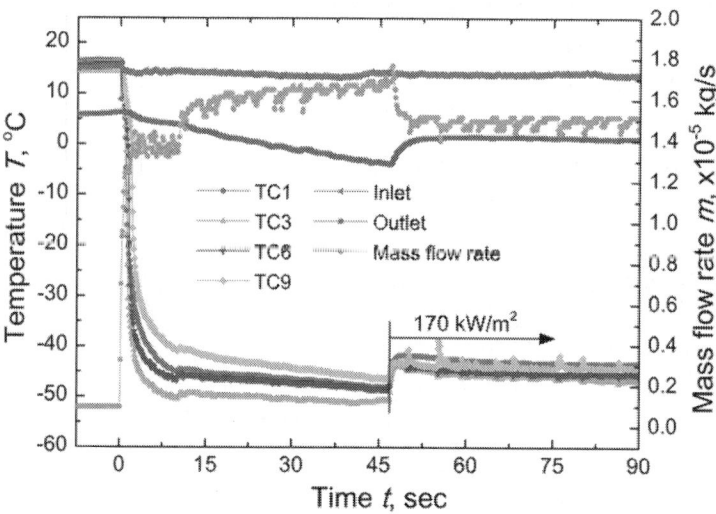

Figure 5: Time variation in temperature at TC1, TC3, TC6, TC9, inlet, and outlet and mass flow rate under heat flux of 170 kW/m².

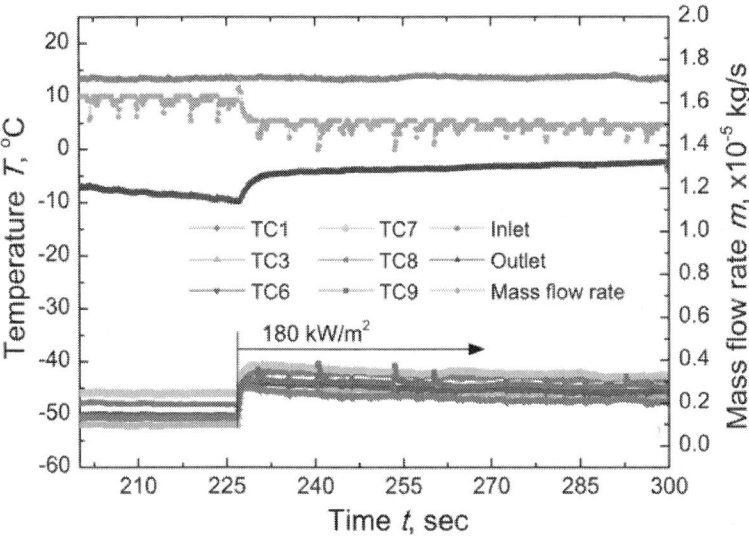

Figure 6: Time variation in temperature at TC1, TC3, TC6, TC7, TC8, TC9, inlet, and outlet, and in mass flow rate under heat flux of 180 kW/m².

HEAT TRANSFER COEFFICIENT

To consider the heat conduction in the channel wall, the one-dimensional axisymmetric heat conduction equation was solved. This equation with volumetric heat generation q_v in the steady state is expressed as

$$\frac{k_s}{r}\frac{\mathrm{d}}{\mathrm{d}r}\left(r\frac{\mathrm{d}T}{\mathrm{d}r}\right)=\dot{q}_v$$

(1)

The boundary conditions can be written as

$$\left.\frac{\mathrm{d}T}{\mathrm{d}r}\right|_{r=r_o}=0$$

(2)

$$-k_s\left.\frac{\mathrm{d}T}{\mathrm{d}r}\right|_{r=r_o}=h\left[T_{sat}-T(r_i)\right]$$

(3)

By solving Equation (1) analytically, the wall superheat and heat transfer coefficient can be expressed as

$$\Delta T_{sat} = T_{surf} - T_{sat} + \frac{q}{k_s}\left(\frac{r_i}{2} - \frac{r_i r_o^2}{r_o^2 - r_i^2}\ln\frac{r_o}{r_i}\right),$$

(4)

$$h = \left[\frac{T_{surf} - T_{sat}}{q} + \frac{1}{k_s}\left(\frac{r_i}{2} - \frac{r_i r_o^2}{r_o^2 - r_i^2}\ln\frac{r_o}{r_i}\right)\right]^{-1},$$

(5)

Where h: heat transfer coefficient [W/(m^2 · K)],k_s: thermal conductivity of stainless steel [W/(m · K)], q: heat flux [W/m^2],r_i: inner radius of outer tube[m] and r_o: outer radius of outer tube [m]. By using Equations (4) and (5), the wall superheat and heat transfer coefficient can be determined from the experimental data.

Figure 8 shows the wall superheat distribution on the cryoprobe surface derived using Equation (4). The wall superheat decreases slightly along the channel. This result indicates that heat conduction through the wall may affect the wall superheat. The heat flux of 150 kW/m^2 shows the highest wall superheat. Figure 9 shows the heat transfer coefficient distribution on the ultrafine cryoprobe surface derived using Equation (5). The heat transfer coefficient obviously increases along the channel, and the location changes dramatically at more than 30 mm. In this region, the axial heat conduction has a strong effect. Therefore, the representative wall superheat and heat transfer coefficient were determined by averaging the value sat locations of less than 30 mm.

Figure 10 shows the relationship between the heat transfer coefficient, wall superheat, and heat flux. Additionally, the estimated heat transfer coefficients are compared with Li's correlation [15] , which is the correlation of the heat transfer coefficient of phase change flow in mini/microchannels, expressed as

$$Nu = 334 Bl^{0.3}\left(Bo \cdot Re_L^{0.36}\right)^{0.4},$$

(6)

Where,

k_L : thermal conductivity of the liquid phase [W/(m · K)],

Nu : Nusselt number,

Bl : boiling number,

Bo : Bond number and

Re_L : Reynolds number of the liquid phase, which are expressed as

$$Nu = \frac{hD}{k_L}, \; Bl = \frac{q}{Gi_{LV}},$$

$$Bo = \frac{g\left(\rho_L - \rho_V\right)D^2}{\sigma}, \; Re_L = \frac{G\left(1 - x_{eq}\right)D}{\mu_L},$$

(7)

Where G: mass flux [kg/(m² · s)], i_{LV}: latent heat [J/kg], g: gravitational acceleration [m/s²]

ρ_L : density of the liquid phase [kg/m³],

ρ_V : density of the vapor phase [kg/m³],

σ : surface tension [N/m],

x_{eq} : vapor quality

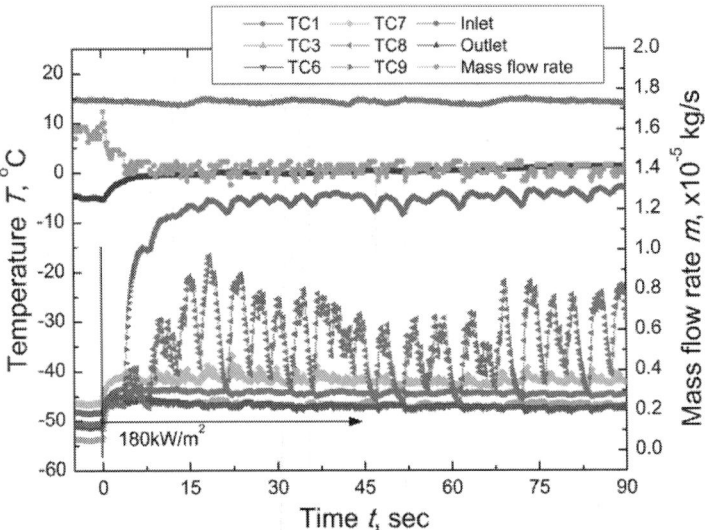

Figure 7: Unstable response in temperature at TC1, TC3, TC6, TC7, TC8, TC9, inlet, and outlet, and in mass flow rate under heat flux of 180 kW/m².

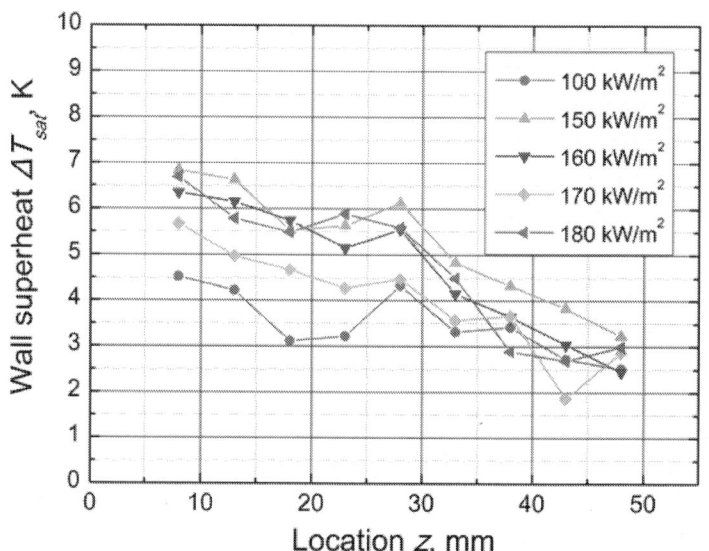

Figure 8: Wall superheat distribution on cryoprobe surface (Location z = 0 represents tip of cryoprobe).

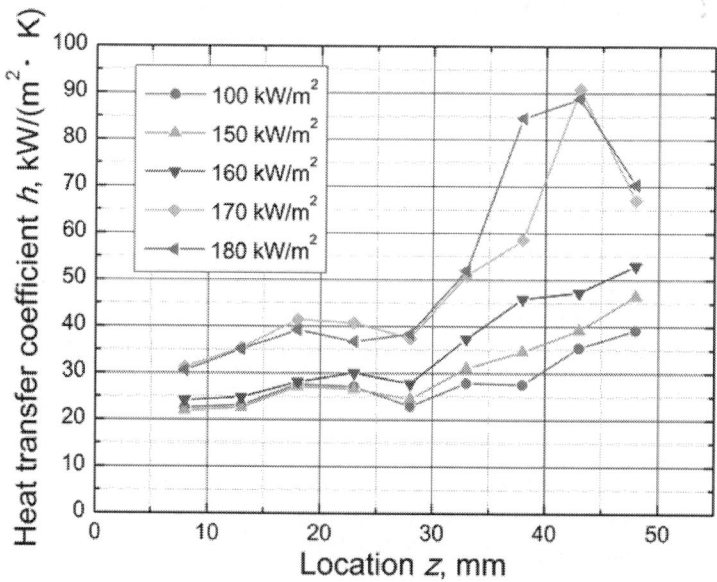

Figure 9: Heat transfer coefficient distribution on cryoprobe surface (Location z = 0 represents tip of cryoprobe).

[-] and μ_L: viscosity [Pa · s]. As shown in Figure 10, the heat transfer coefficient reaches a maximum value of 39 kW/(m² · K) at

a heat flux of 170 kW/m². Furthermore, the heat transfer coefficient increases dramatically when the heat flux exceeds 150 kW/m². Compared with the value from Li's correlation, the experimental value was close to the heat transfer coefficient for low vapor quality under a low heat flux. However, in the high heat flux region, the rate of increase of the experimental results was larger than that from Li's correlation. This strong dependence on the heat flux is a general tendency of nucleate boiling. Therefore, Figure 10 implies that the heat transfer mode in the ultrafine cryoprobe is nucleate boiling.

MODE OF PHASE CHANGE HEAT TRANSFER IN CRYOPROBE

To evaluate the phase change heat transfer regime, the bubble expansion model [16] was applied to the ultrafine cryoprobe, as shown in Figure 11. Here, the bubble is assumed to travel at the velocity of the liquid phase and to expand only downstream. In addition, the annular channel of the ultrafine cryoprobe is assumed to be the equivalent circular channel by using the hydraulic diameter.

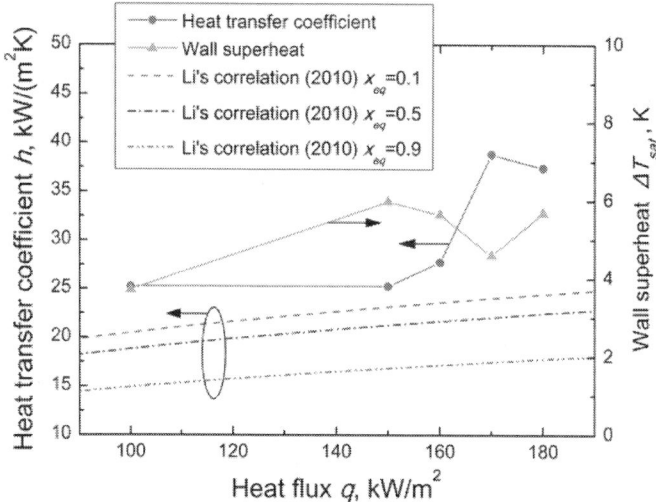

Figure 10: Relationship between heat transfer coefficient, wall superheat, and heat flux.

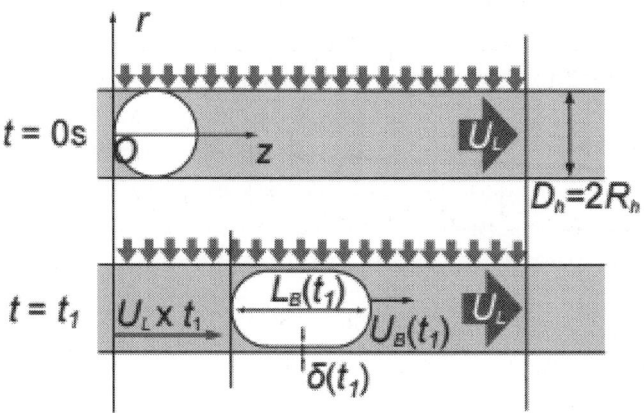

Figure 11: Bubble expansion model.

The energy balance between the bubble volume and supplied heat is expressed as

$$\rho_V i_{LV} \frac{dV_B(t)}{dt} = \pi D L_B(t) q$$

(8)

Where V_B: bubble volume [m³], D: channel diameter [m], L_B: bubble length [m]. The bubble head speed U_B is defined as the time derivative of the bubble length as follows:

$$U_B(t) = \frac{dL_B(t)}{dt}$$

(9)

The liquid film thickness is determined by Taylor's law, which is the correlation between the liquid film thickness and bubble velocity, expressed as

$$\frac{2\delta}{D} = \frac{0.643(3Ca)^{2/3}}{1+1.608(3Ca)^{2/3}}$$

(10)

Where, δ : liquid film thickness [m],
$Ca = \mu_L U_B / \sigma$: capillary number.

This model can calculate the bubble expansion process and the temporal variation in the liquid film thickness during expansion. The heat transfer coefficient can be calculated from the liquid film thickness with following equation as h = $k_{L/\delta}$.

Figure 12 compares the experimentally obtained heat transfer coefficient with the theoretical prediction of the bubble expansion model. The theoretical prediction underestimates the experimental value. Additionally, Figure 13 compares the wall superheat from the experiment with the theoretical prediction of the bubble expansion

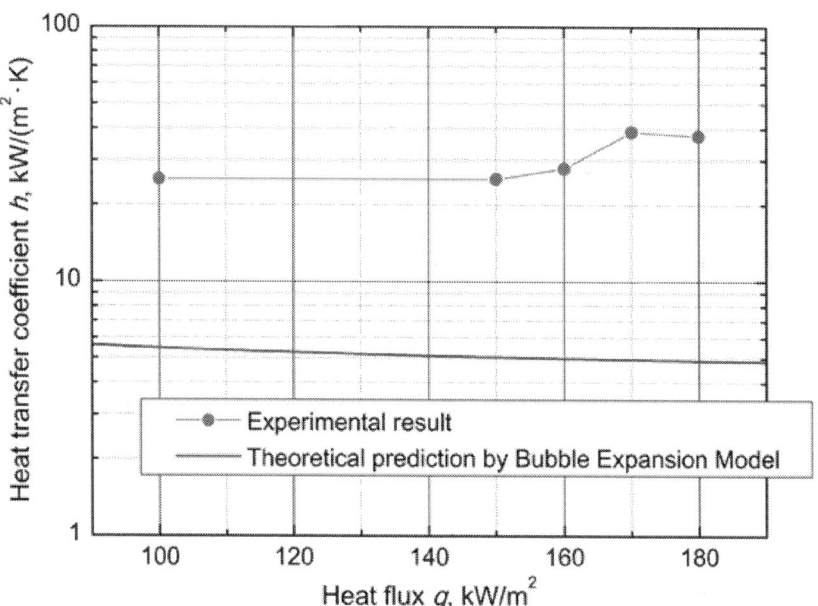

Figure 12: Comparison of heat transfer coefficients estimated from experimental data and from bubble expansion model.

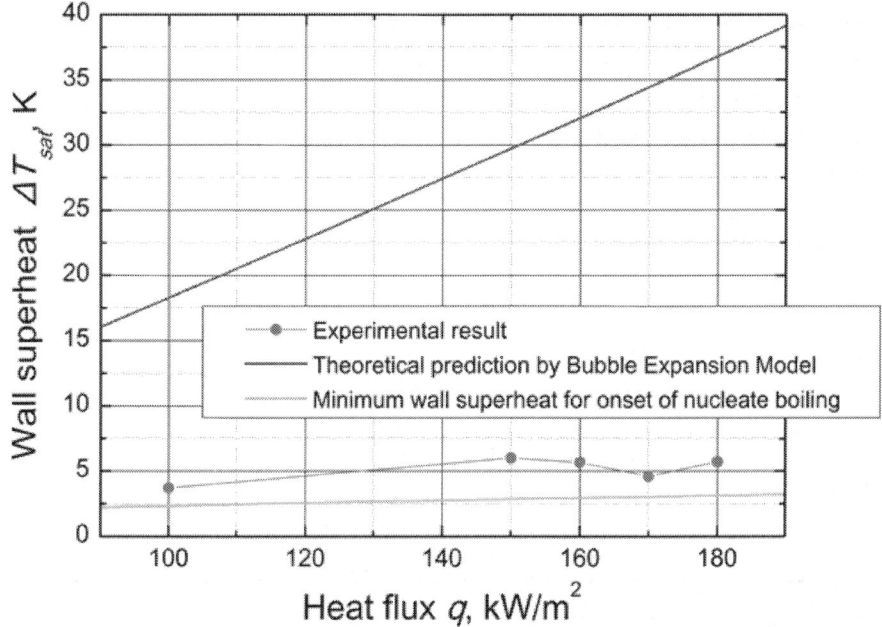

Figure 13: Comparison of wall superheat estimated from experimental data and that estimated from bubble expansion model.

model. The experimental results are much smaller than the theoretical prediction. In addition, the experimental results are larger than the minimum wall superheat for the onset of nucleate boiling [16] . Nucleate boiling does not occur if the wall superheat is less than this minimum value. Actually, it is difficult to measure the size and distribution of the cavity. Therefore, the minimum wall superheat to produce nucleate boiling is an important parameter. This parameter is determined by the relationship between the wall heat flux q_w and the wall superheat ΔT_{sat} to initiate heterogeneous nucleation, which is expressed as [17]

$$\Delta T_{sat} = \frac{\dfrac{R_g T_{sat}^2}{i_{LV}} \ln\left(1 + \dfrac{2\sigma}{p_L r_a}\right)}{1 - \dfrac{R_g T_{sat}}{i_{LV}} \ln\left(1 + \dfrac{2\sigma}{p_L r_a}\right)} + \frac{q r_a}{k_L} \quad , \tag{11}$$

whereR_g: gas constatnt [J/(kg · K)],r_a: active cavity size [m], and P_L: pressure in the liquid phase [Pa]. Therefore, the liquid film thickness predicted by the bubble expansion model is too thick to maintain a low wall temperature. Hence, the heat transfer mode in the ultrafine cryoprobe is not pure evaporative heat transfer from the liquid film. These results may indicate that phase change heat transfer in the ultrafine cryoprobe occurs by nucleate boiling.

CONCLUSIONS

In this study, the characteristics of phase change heat transfer in an ultrafine cryoprobe were evaluated. The heat transfer coefficient obtained from the experimental data was compared with that obtained from the conventional correlation and the bubble expansion model. The results obtained in this study can be summarized as follows:

1) The ultrafine cryoprobe cooled by HFC-23 exhibited a stable surface temperature of −50°C.
2) The heat transfer coefficient of the ultrafine cryoprobe was estimated. It reached a maximum value of 39 kW/(m² · K) at a heat flux of 170 kW/m².
3) By using the bubble expansion model, the phase change heat transfer regime in the ultrafine cryoprobe was estimated. The results may indicate that nucleate boiling occurs in the cryoprobe.
4) By using the developed cryoprobe, the small-scale cryosurgery will be achieved. We can take advantage of the small size of cryoprobe to treat small lesions, such as pigments on the skin, wrinkles around the eyes, and early breast cancer, with minimal invasiveness. Additionally, the cryosurgery in the blood vessel combined with the catheter is a possible application in the future.

ACKNOWLEDGEMENTS

J. Okajima received support from a Grant-in-Aid for Young Scientists (B) [25820054] from the Japan Society for the Promotion of Science.

REFERENCES

1. Aihara, T., Kim, J.-K., Suzuki, K. and Kasahara, K. (1993) Boiling Heat Transfer of a Micro-Impinging Jet of Liquid Nitrogen in a Very Slender Cryoprobe. International Journal of Heat and Mass Transfer, 36, 169-175. http://dx.doi.org/10.1016/0017-9310(93)80076-7
2. Bénita, M. and Condé, H. (1972) Effects of Local Cooling upon Conduction and Synaptic Transmission. Brain Research, 36, 133-151. http://dx.doi.org/10.1016/0006-8993(72)90771-8
3. Bischof, J., Christov, K. and Rubinsky, B. (1993) A Morphological-Study of Cooling Rate Response in Normal and Neoplastic Human Liver-Tissue-Cryosurgical Implications. Cryobiology, 30, 482-492. http://dx.doi.org/10.1006/cryo. 1993.1049
4. Davis, E.J. and Anderson, G.H. (1966) The Incipience of Nucleate Boiling in Forced Convection Flow. AIChE Journal, 12, 774-780. http://dx.doi.org/10.1002/aic.690120426
5. Doll, N., Meyer, R., Walther, T. and Mohr, F.W. (2004) A New Cryoprobe for Intraoperative Ablation of Atrial Fibrillation. The Annals of Thoracic Surgery, 77, 1460-1462.http://dx.doi.org/ 10.1016/S0003-4975(03)01389-4
6. Li, W. and Wu, Z. (2010) A General Correlation for Evaporative Heat Transfer in Micro/Mini-Channels. International Journal of Heat and Mass Transfer, 53, 1778-1787.http://dx.doi.org/ 10.1016/j.ijheatmasstransfer.2010.01.012
7. Maruyama, S., Nakagawa, K., Takeda, H., Aiba, S. and Komiya, A. (2008) The Flexible Cryoprobe Using Peltier Effect for Heat Transfer Control. Journal of Biomechanical Science and Engineering, 3, 138-150. http://dx.doi.org/10.1299/jbse.3.138

8. Okajima, J., Komiya, A. and Maruyama, S. (2010) Boiling Heat Transfer in Small Channel for Development of Ultrafine Cryoprobe. International Journal of Heat and Fluid Flow, 31, 1012-1018. http://dx.doi.org/10.1016/j.ijheatfluidflow.2010.08.008

9. Okajima, J., Komiya, A. and Maruyama, S. (2012) Analysis of Evaporative Heat Transfer by Expansion Bubble in a Microchannel for High Heat Flux Cooling. Journal of Thermal Science and Technology, 7, 740-752. http://dx.doi.org/10.1299/jtst.7.740

10. Okajima, J., Komiya, A. and Maruyama, S. (2013) 24-Gauge Ultrafine Cryoprobe with Diameter of 550 µm and Its Cooling Performance. Submitted to Cryobiology.

11. Okajima, J., Maruyama, S., Takeda, H., Komiya, A. and Sangkwon, J. (2010) Cooling Characteristics of Ultrafine Cryoprobe Utilizing Convective Boiling Heat Transfer in Microchannel. Proceedings of 14th International Heat Transfer Conference, Washington DC, 8-13 August 2010, 297-306. http://dx.doi.org/10.1115/IHTC14-22550

12. Popken, F., Land, M., Bosse, M., Erberich, H., Meschede, P., Konig, D.P., Fischer, J.H. and Eysel, P. (2003) Cryosurgery in Long Bones with New Miniature Cryoprobe: An Experimental in Vivo Study of the Cryosurgical Temperature Field in Sheep. European Journal of Surgical Oncology, 29, 542-547. http://dx.doi.org/10.1016/S0748-7983(03)00069-6

13. Rewcastle, J.C., Sandison, G.A., Saliken, J.C., Donnelly, B.J. and McKinnon, J.G. (1999) Considerations during Clinical Operation of Two Commercially Available Cryomachines. Journal of Surgical Oncology, 71, 106-111. http://dx.doi.org/ 10.1002/(SICI)1096-9098(199906)71:2<106::AID-JSO9>3.0.CO;2-Z

14. Seifert, J.K., Gerharz, C.D., Mattes, F., Nassir, F., Fachinger, K., Beil, C. and Junginger, T. (2003) A Pig Model of Hepatic Cryotherapy. In Vivo Temperature Distribution during Freezing and Histopathological Changes. Cryobiology, 47, 214-226. http://dx.doi.org/10.1016/j.cryobiol.2003.10.007

15. Tacke, J., Adam, G., Haage, P., Sellhaus, B., Großkortenhaus, S. and Günther, R.W. (2001) MR-Guided Percutaneous Cryotherapy of the Liver: In Vivo Evaluation with Histologic Correlation in an Animal Model. Journal of Magnetic Resonance

Imaging, 13, 50-56. http://dx.doi.org/10.1002/1522-2586 (200101) 13:1<50::AID-JMRI1008>3.0.CO;2-A

16. Takeda, H., Maruyama, S., Okajima, J., Aiba, S. and Komiya, A. (2009) Development and Estimation of a Novel Cryoprobe Utilizing the Peltier Effect for Precise and Safe Cryosurgery. Cryobiology, 59, 275-284. http://dx.doi.org/10.1016/j.cryobiol.2009.08.004

17. Zhang, J.-X., Ni, H. and Harper, R.M. (1986) A Miniaturized Cryoprobe for Functional Neuronal Blockade in Freely Moving Animals. Journal of Neuroscience Methods, 16, 79-87.http://dx.doi.org/10.1016/0165-0270(86)90010-5

NOMENCLATURE

D : Channel diameter [m]

G : Mass flux [kg/(m^2 · s)]

g : Gravitational acceleration [m/s^2]

h : Heat transfer coefficient [W/(m^2 · K)]

i_{LV} : Latent heat [J/kg]

k : Thermal conductivity [W/(m · K)]

L_B : Bubble length [m]

p : Pressure [Pa]

q : Heat flux [W/m^2]

$\dot{q}_v$: Volumetric heat generation [W/m^3]

R_g : Gas constatnt [J/(kg · K)]

r : Radius, coordinate on radial direction [m]

r_a : Active cavity size [m]

U_B : Bubble head speed [m/s]

V_B : Bubble volume [m³]

x_{eq} : Vapor quality [-]

z : Coordinate on axial direction [m]

Greek

δ : liquid film thickness [m]

μ : viscosity [Pa · s]

ρ : density of the liquid phase [kg/m³]

σ : surface tension [N/m]

Dimensionless number

Bl : Boiling number

Bo : Bond number

Ca : Capillary number

Nu : Nusselt number

Re_L : Reynolds number of the liquid phase

Subscript

i : Inner tube

L : Liquid phase

o : Outer tube

s : Stainless steel

CITATION

Junnosuke Okajima, Atsuki Komiya, Shigenao Maruyama, Evaluation of Characteristics of Phase Change Heat Transfer in Ultrafine Cryoprobe, Journal of Flow Control, Measurement & Visualization Vol.02 No.02(2014), Article ID:44769,11 pages 10.4236/jfcmv.2014.22008

CHAPTER 10

Heat Transfer Characteristics of Square Micro Pin Fins Under Natural Convection

Naoko Matsumoto, Toshio Tomimura, Yasushi Koito

Department of Advanced Mechanical Systems, Kumamoto University, Kumamoto, Japan

ABSTRACT

In order to comply with the recent demand for downsizing of the electric equipment, the minia- turization and the improvement in heat transfer performance of a heat sink under natural air-cooling are increasingly required. This paper describes the experimental and numerical investigations of heat sinks with miniature/micro pins and the effect of the pin size, pin height and the number of pins on heat transfer characteristics of heat sinks. Five types of basic heat sink models are investigated in this study. The whole heat transfer area of heat sinks having the different pin size, pin height and the number of pins respectively is kept constant. From a series of experiments and numerical analyses, it has been clarified that the heat sink temperature rises with increase in the number of pins. That is, the heat sink with miniaturized fine pins showed almost no effect on the heat transfer enhancement. This is because of the choking phenomenon occurred in the air space among the pin fins. Reflecting these results, it is confirmed that the heat transfer coefficient reduces with miniaturization of pins. Concerning the effects of the heat transfer area on the heat sink performance, almost the same tendency has been observed in other three series of large surface area, that is, higher pin height. Furthermore as a result of studying non-dimensional convection heat transfer performance, it was found that the relation between the Nusselt number (Nu) and the Rayleight number (Ra) is given by $Nu = 0.16\, Ra^{0.52}$.

Keywords
Natural Air-Cooling, Heat Sink, Micro Pin Fin, Heat Transfer Performance, Experiment, Numerical Analysis

INTRODUCTION

In electronic equipment applications with high heat fluxes, the heat transfer performance and the lifetime of components are often deteriorated due to temperature rise. To avoid those unfavorable problems, it is necessary to keep the component under a critical temperature by introducing a high performance heat sink. Furthermore, in order to comply with the recent demand for downsizing of the electronic equipment, the miniaturization and the improvement in heat transfer performance of the heat sink are required much more strongly than ever. But it is well known that the decrease in heat transfer area by downsizing leads to a decline of the heat sink performance.

Many studies of heat transfer characteristics using large scaled heat sinks have been carried out so far. For example, Aihara et al.

[1] and Sara [2] revealed that the heat sink performance depends on the pin size, the population density of pin. Zografos and Sunderland [3] studied the effect of pin arrangement on the heat transfer performance, and reported that a choking phenomenon, which had a negative effect on heat transfer from a heat sink, was observed in the air space among pin fins. Huang et al. [4] , Sparrow et al. [5] and Sertkaya et al. [6] investigated experimentally the dependence of the pin fin performance on a heat sink orientation.

In electronic equipment applications, heat sinks are placed within enclosed areas and the emphasis is on making the enclosures more compact than ever. Yu et al. [7] and Bocu et al. [8] reported natural convection heat transfer from heat sink attached to enclosures in their studies.

On the other hand, concerning small sized heat sinks or nano particles structure, the analytical and experimental studies were published by Narasimhan et al. [9] and Minakami et al. [10] , Kunugi et al. [11] . Although interesting findings on natural and forced convection, and radiation from heat sinks have been reported in the previous studies, the mechanism of the heat transfer has not been clarified yet.

Accordingly, in the present study, the effects of pin size, pin height and the number of pins on the heat transfer characteristics of heat sinks with miniature/micro pins have been investigated fundamentally.

HEAT SINK MODELS

Five types of basic heat sink models are shown in Figure 1. The first model has four square pins, and is named as Type 1. Each pin placed at equal space of the heat sink base (length: 25 mm, width: 25 mm, thickness: 2 mm) has the pin width w_p = 6.25 mm and the pin height h_p = 6.5, 14, 21.5, 29 mm at intervals of 7.5 mm. And in Type 2 model, the pin size is reduced to half the pin width and height of Type 1. That is, under the conditions of constant heat transfer area, the number of pins for Type 2 is quadrupled against Type 1. In the same manner, five types of heat sinks in total, which have the pin height from several hundred micro meters to a few millimetres and have the same heat transfer area, are prepared.

As listed in Table 1, the number of pins N is changed from 4 to 1024, and corresponding to this change in N, the pin width w_p and the pin height h_p vary from 0.39 mm to 6.25 mm and from 0.41 mm to 29 mm, respectively. The whole heat transfer area of the square pin series A_{hs} is 1.48 ´ 10^{-3} m², 2.23 ´ 10^{-3} m², 2.98 ´ 10^{-3} m² and 3.73 ´ 10^{-3} m².

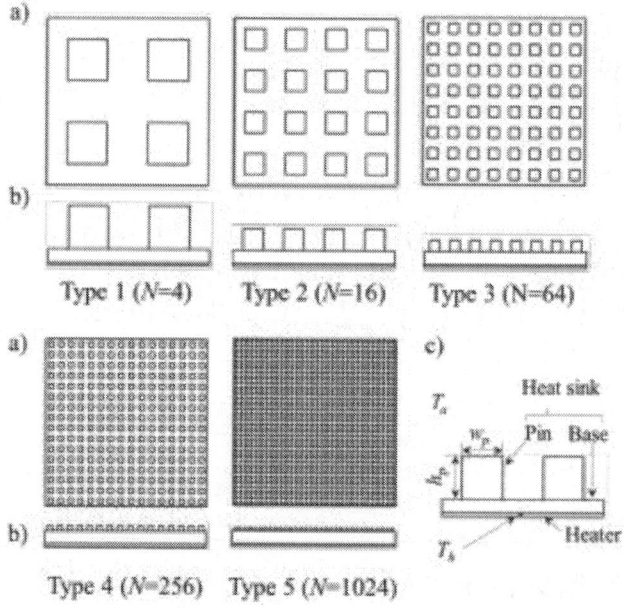

Figure 1: Heat sink models, (a) Top view, (b) Side view, (c) Parts and nomenclature.

Table 1: Dimensions of pin fin.

Type	N (–)	w_p (mm)	h_p (mm)			
			●	□	⎯	△
1	4	6.25	6.5	14	21.5	29
2	16	3.13	3.25	7	10.75	14.5
3	64	1.56	1.63	3.5	5.38	7.25
4	256	0.78	0.81	1.75	2.69	3.63
5	1024	0.39	0.41	0.88	1.34	1.81

EXPERIMENTAL

Figure 2 shows a schematic diagram of experimental apparatus, which consists of a chamber, a vacuum pump, a vacuum gauge, and a heat sink with a heater used as the heat source. The chamber made of SUS is 400 mm in diameter and 170 mm in height. Here, its wall temperature which is regarded as the ambient temperature T_a was kept at 25°C by circulating cooling water. The heat sink was hung in the air in the chamber by using fine wires to reduce the heat loss from the heat sink to the chamber wall by conduction. The heat sink was placed in the upward facings of the heated surface. The pressure in the chamber was controlled with the vacuum pump from the atmospheric pressure to the vacuum condition of approximately 3.5×10^{-2} Pa, and the chamber pressure was measured using the vacuum gauge. Under the atmospheric pressure, heat is transferred from the heat sink to the surroundings by convection and thermal radiation. On the other hand, under the vacuum condition of approximately 3.5×10^{-2} Pa, heat is transferred from the heat sink to the surrounding wall only by thermal radiation.

As shown in Figure 3, the heat sink, which is made of Aluminum alloy A6063 (the thermal conductivity l = 200 W/(m·K), the emissivity e = 0.89) with a base plate of 25 mm long, 25 mm wide and 2 mm thick, was placed on the same sized square copper block (thickness: 2 mm, l = 398 W/(m·K)) through a thermal interface material (thickness: 0.5 mm, l = 3.1 W/(m·K)). Furthermore, on the bottom part of the copper block, a ceramic heater is attached through the thermal interface material. The representative temperature T_h was measured using a f 0.5 mm K-type thermocouple fixed in a small groove machined at the copper block center. In Figure 3, T_a is the ambient temperature.

First, under the atmospheric pressure, the heat rate Q_{in} = 1.25 W was applied to the heat sink by the ceramic heater. Second, the representative temperature T_h was measured under the steady-state condition. Third, under the vacuum condition of approximately 3.5 × 10^{-2} Pa, some amount of heat was applied to the heat sink by the heater until the heat sink temperature became the same temperature T_h under the atmospheric pressure, and then the heat transferred by thermal radiation Q_r from the heat sink under the steady-state conditions was measured based on the input power to the heater. Furthermore, in the same manner, T_h and Q_r for the heat sink consisting only of the base plate were also measured for reference.

NUMERICAL ANALYSIS

Figure 4 shows the boundary conditions for the large air space surrounding the heat sink placed in the upward facings of the heated surface. A series of numerical analyses have been performed for the physical models shown in Figure 1 and for the heat sink consisting only of the base plate (25 mm, 50 mm, 100 mm and 150 mm on a side) for reference. In this study, the fine wires for hanging the heat sink and the copper block, the thermal interface material were neglected for simplification, and only the heater (length: 25 mm, width: 25 mm, thickness: 0.5 mm), the heat sink, and the surrounding large air space were taken into consideration.

The present numerical calculations were carried out using the simulation software of Computational Fluid Dynamics (CFD). The following assumptions are applied in the analysis.

1) Fluid is incompressible.
2) Thermophysical properties are assumed to be constant except density (Boussinesq approximation).
3) Ambient temperature T_a is constant.
4) Based on these assumptions, the following basic equations are obtained under the steady state conditions.

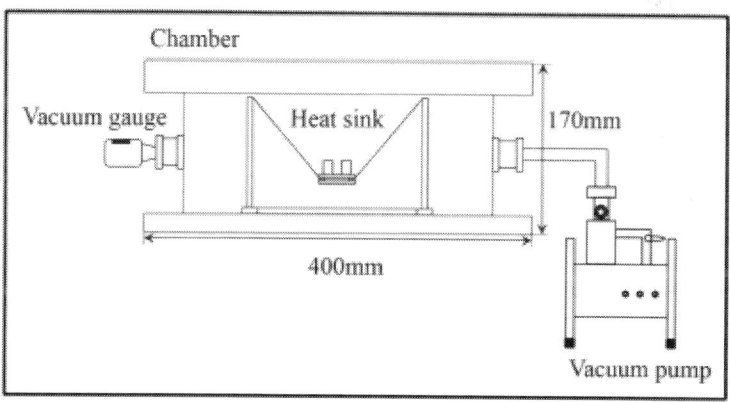

Figure 2: Schematic of experimental apparatus.

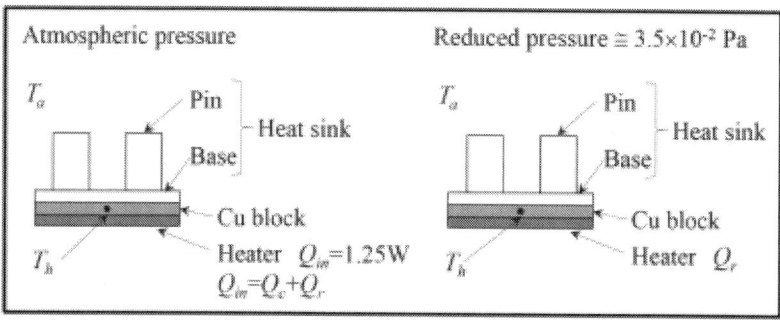

Figure 3: Configuration of tested heat sink.

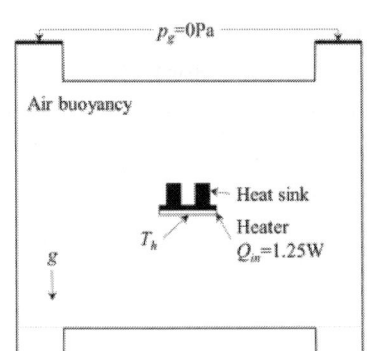

Figure 4: Boundary conditions of CFD (upward facing).

Continuity equation:

$$\frac{\partial u}{\partial x}+\frac{\partial v}{\partial y}+\frac{\partial w}{\partial z}=0$$

(1)

Momentum equations:

$$u\frac{\partial u}{\partial x}+v\frac{\partial u}{\partial y}+w\frac{\partial u}{\partial z}=-\frac{1}{\rho}\frac{\partial p}{\partial x}+\upsilon\left(\frac{\partial^{2}u}{\partial x^{2}}+\frac{\partial^{2}u}{\partial y^{2}}+\frac{\partial^{2}u}{\partial z^{2}}\right)\text{(x-direction)}$$

(2)

$$u\frac{\partial v}{\partial x}+v\frac{\partial v}{\partial y}+w\frac{\partial v}{\partial z}=-\frac{1}{\rho}\frac{\partial p}{\partial y}+\upsilon\left(\frac{\partial^{2}v}{\partial x^{2}}+\frac{\partial^{2}v}{\partial y^{2}}+\frac{\partial^{2}v}{\partial z^{2}}\right)\text{(y-direction)}$$

(3)

$$u\frac{\partial w}{\partial x}+v\frac{\partial w}{\partial y}+w\frac{\partial w}{\partial z}=-\frac{1}{\rho}\frac{\partial p}{\partial z}+\upsilon\left(\frac{\partial^{2}w}{\partial x^{2}}+\frac{\partial^{2}w}{\partial y^{2}}+\frac{\partial^{2}w}{\partial z^{2}}\right)+g\beta(T-T_{a})$$

(z-direction)

(4)

Energy equation:

$$u\frac{\partial T}{\partial x}+v\frac{\partial T}{\partial y}+w\frac{\partial T}{\partial z}=\alpha\left(\frac{\partial^{2}T}{\partial x^{2}}+\frac{\partial^{2}T}{\partial y^{2}}+\frac{\partial^{2}T}{\partial z^{2}}\right)$$

(5)

Boundary conditions are given as follows. The same heat rate Q_{in} = 1.25 W as the experiment was applied to the heat sink. The velocity is zero on all the solid surfaces, that is u = v = w = 0. The gauge pressure p_g of the inlet and the outlet of the large surrounding space is 0 Pa, the inlet air temperature is 25°C, and other remaining surfaces are adiabatic. The amount of heat transfer rate by convection and thermal radiation is equal to that by conduction through the solid.

The equation of conduction within the solid is given by the following equation.

$$\frac{\partial^2 T}{\partial x^2} + \frac{\partial^2 T}{\partial y^2} + \frac{\partial^2 T}{\partial z^2} = 0$$

(6)

In the numerical analysis, the same heat rate Q_{in} = 1.25 W as the experiment was applied to the heater. In the case of large sized heat sink consisting only of the base plate, the heat rate added to the base plate was kept constant.

Then the representative temperature T_h of the heat sink under the steady-state conditions and the heat rate released from the heat sink by thermal radiation Q_r were calculated using the simulation software of CFD. In addition, the velocity vectors and the temperature fields at the central lateral plane of the heat sink were obtained.

RESULTS AND DISCUSSION

Heat Transfer Performance of Heat Sinks

Figure 5 and Figure 6 show the measured and simulated results on the effects of the number of pins N, the pin height h_p on the temperature rise DT of the heat sink. Here, DT is defined as the difference between the repre-

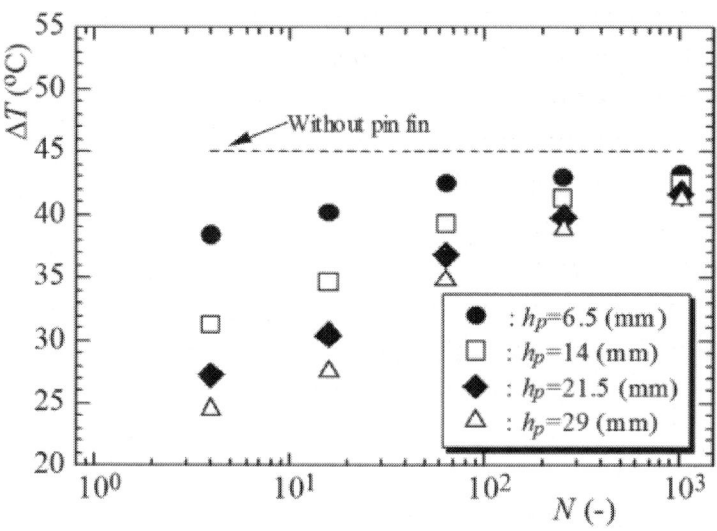

Figure 5: Effects of N and h_p on DT (measured results).

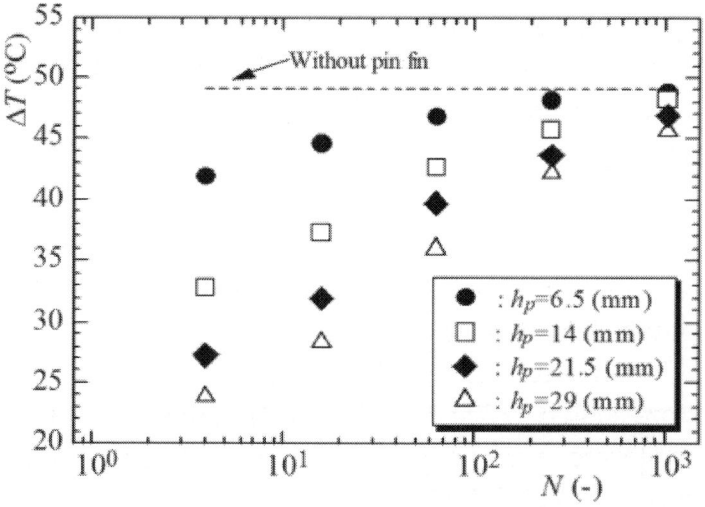

Figure 6: Effects of N and h_p on DT (calculated results).

sentative temperature T_h and the ambient temperature T_a. Accordingly, the lower DT means a higher heat transfer performance. The symbols show the results for the heat sink with different pin height, that is, heat transfer area. The broken line shows the DT for the heat sink consisting only of the base plate. Further, the number of pins N corresponds to the heat sink type from Type 1 to Type 5.

Concerning Type 1 model having the smallest number of pins, it is found that the heat sink performance of Type 1 model is the highest and as the pin height h_p is higher, the heat sink performance is better. In a series of heat sinks, the heat transfer rate increases with the pin height of Type 1. Each representative temperature T_h of the heat sink rises with increase in the number of pins N. Especially, in the case of large N, the temperature rise DT of the heat sink comes close to that of the base plate level. In other words, the heat sink with miniaturized pins has almost no effect on the heat transfer enhancement. As seen from the figure, the simulated results agree qualitatively well with the measured results. That is, the dependence of the temperature rise DT on the number of pins N agrees fairly well with each other. Unfortunately, however, a quantitative agreement is not obtained between them. This is because the present physical model cannot reproduce the experimental heat sink system exactly.

Velocity Vector and Temperature Field

The velocity vector in air space surrounding the heat sink model at central longitudinal section, which are obtained from numerical simulation, are shown in Figure 7. From this figure, it can be seen that the velocity vector among pins becomes small with increase in the number of pins. In particular, the air velocity of Type 5 models becomes almost zero.

The temperature fields of the heat sink and its surrounding air space are shown in Figure 8. The representative temperature T_h at the heater surface is the highest. From this figure, it can be seen that the heater tem- perature rises from Type 1 to Type 5. Namely, this means that the increase in the number of pins causes the deterioration of the heat sink performance.

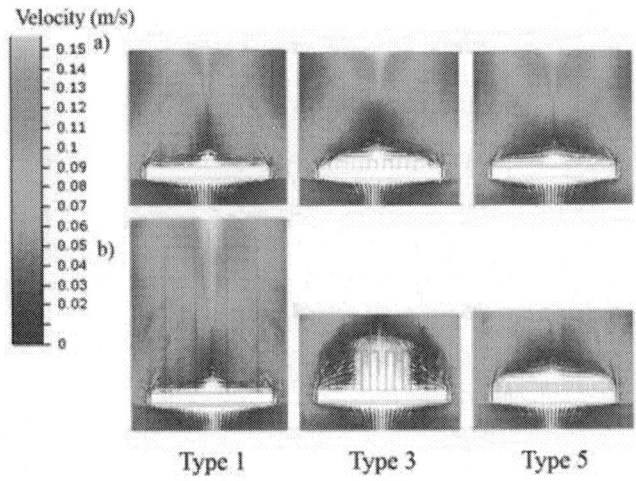

Figure 7: Flow field along center line of heat sink; (a) $h_p = 6.5$ mm, (b) $h_p = 29$ mm series.

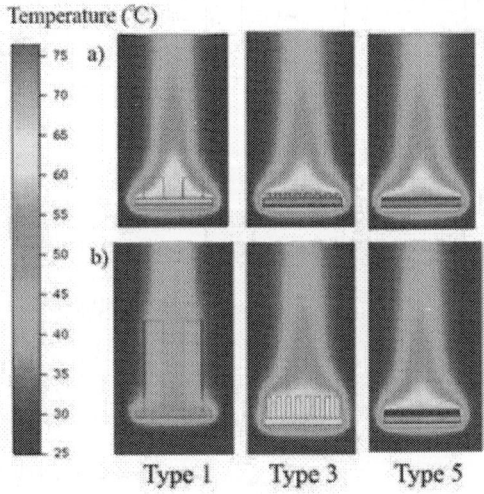

Figure 8: Temperature field along center line of heat sink; (a) $h_p = 6.5$ mm, (b) $h_p = 29$ mm series.

Air Velocity between Pins

Figure 9 shows the air velocity between pins which was numerically obtained for Type 1 to Type 5 models. The present air velocity means the average value obtained at central lateral plane of heat sink. As shown in the figure, the air velocity decreases with increase in the number of pins. This result shows the same tendency as the velocity vectors. In Type 3 to Type 5 models, regardless of the pin height, the air velocity becomes almost zero. This behavior must be the choking phenomena observed in the previous study using the large sized heat sinks done by Zografos et al. [3] , and as an inevitable consequence, it leads to low heat transfer performance of the heat sink.

The Ratio of Convective and Radiative Heat Transfer Rates

In order to estimate the heat sink performance from the view point of the heat transfer, the ratio of convective and radiative heat transfer rates in whole heat sink system has been also investigated experimentally and analytically, and the measured and calculated results are shown in Figure 10and Figure 11 respectively. The total amount of heat transfer rate into and out has to keep the energy balance expressed by Equation (7). Under the conditions of the experiments and numerical calculations, Q_{in} is 1.25 W. In addition, the output energy from heat sink is the sum of convective and radiative heat transfer rates Q_c and Q_r, so the amount of heat transferred by convection Q_c is given by the following Equation (8).

$$Q_{in} = Q_{out} \tag{7}$$

$$Q_c = Q_{in} - Q_r \tag{8}$$

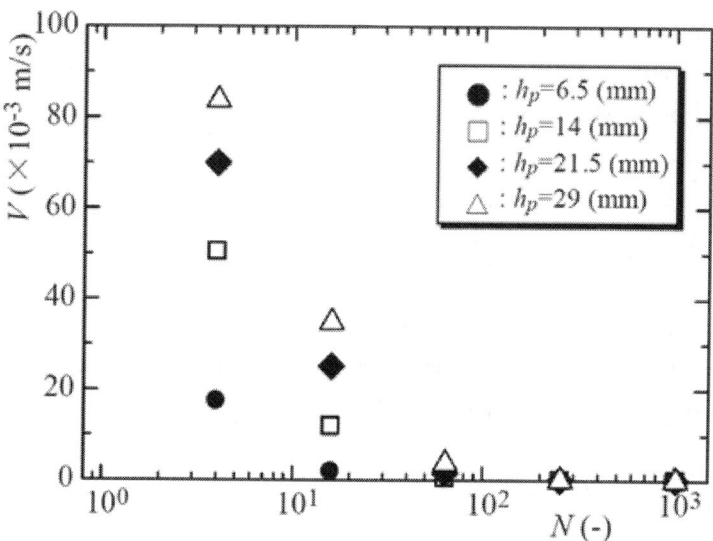

Figure 9: Flow velocity between pins on heat sink.

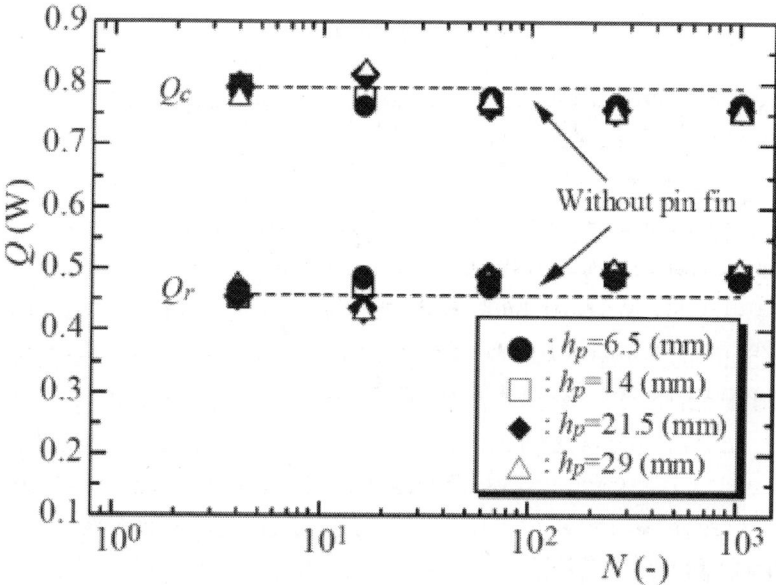

Figure 10: Effects of N and h_p on Q (measured results).

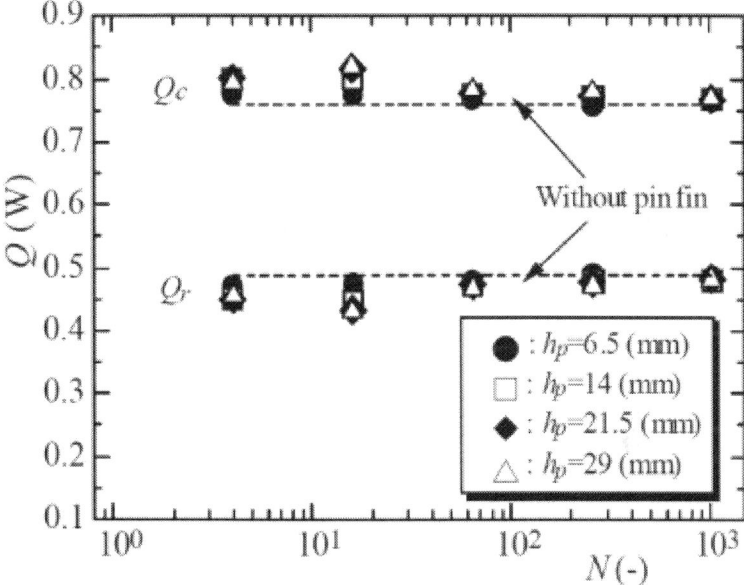

Figure 11: Effects of N and h_p on Q (calculated results).

As seen from Figure 10, it is clear that the convective component Q_c which is the dominant parameter in this case attains its maximum by Type 2, and reduces to Type 5 and then becomes saturated. It can be observed that the convection component Q_c of Type 2 is exactly proportional to the pin height h_p, in other words the heat rate transferred by thermal radiation Q_r vary inversely proportional to the pin height h_p. And in other heat sink models there is no correlation between the convection component Q_c and the pin height h_p. Furthermore, the heat rate transferred by convection Q_c occupies approximately 60% of the total input or output energy Q_{in} or Q_{out}, the heat rate transferred by thermal radiation Q_r is about 40%, and the result agrees very well with the results reported by Sparrow et al. [5] . And from Figure 11, it is found that the calculated results agree well with the measured results qualitatively. However a quantitative agreement is not obtained between them, as previously noted, this is because the present physical model cannot reproduce the experimental heat sink system exactly.

Heat Transfer of Heat Sink under Natural Convection

In numerical simulations, the total amount of heat transfer rates of the heater and the heat sink was investigated. The convective component of heat sink only is represented by Equation (9) using computable parameters of the total amount of convective heat transfer rate Q_c and the heat transfer from heater by convection Q_{c_ht}.

$$Q_{c\_hs} = Q_c - Q_{c\_ht} \tag{9}$$

The convective heat transfer rate of the heater is calculated by Equation (10),

$$Q_c = \frac{Nu\lambda_a}{L} A(T_h - T_a) \tag{10}$$

$$h = \frac{Nu\lambda_a}{L} \tag{11}$$

where Nu is the Nusselt number, l_a is thermal conductivity of air, L is characteristic dimension and A is heat transfer surface of the heater. Equation (11) gives the heat transfer coefficient h.

It is found that the heater has isothermal heating surface according to the simulated results, the Nusselt numbers at the lower and side surfaces of the heater are calculated by previously-proposed Equation (12) [12] , Equation (13) [13] and Equation (14), where Nu_1 is the Nusselt number of lower surface of heated plate, Nu_2 is that of vertical surface of heated plate. And Ra is the Rayleigh number which is represented as the correlation parameter between the Grashof number and the Prandtl number.

$$Nu_1 = 0.27Ra^{0.25} \tag{12}$$

$$Nu_2 = 0.59Ra^{0.25}$$

(13)

where

$$Ra = GrPr = \frac{g\beta(T_h - T_a)L^3}{\upsilon} \frac{\upsilon \rho c_p}{\lambda_a}$$

(14)

Two types of characteristic lengths L will be employed here. In the case of the lower surface of cooled plate, L is taken as L = A/P where A is the plate surface area and P is the perimeter which surrounds the area [14] . In the vertical flat plate, L is taken as 0.5 mm which is the height of heater. Using these numbers, Q_{c_ht} of the each side is calculated by substituting Nu_1 and Nu_2 into Equation (10). And then Q_{c_hs} is determined by Equation (9) using known Q_c.

Figure 12 shows the effects of the number of pins and the pin height on the convective heat transfer coefficient. Here, in order to evaluate the improvement in the convective heat transfer performance for heat sink base surface, the surface area of the top of the base is adopted as A (6.25×10^{-4} m^2 constant). In this figure, it is clear that each heat transfer coefficient h of the heat sink decreases with increase in the number of pins N and in the case of Type 5 model it comes close to that of the base plate level. Since Figure 12 agrees well with the results of the temperature rise, it is considered that the factor of decreasing heat transfer performance due to the miniaturization is the convective heat transfer coefficient.

Finally, the calculated data of heat sink, expressed in terms of Nu and Ra, are plotted in Figure 13. In this figure, the lower solid line represents Equation (15) [14] in which Nu_3 is the Nusselt number of upper surface of heated plate and the symbols (´) show the calculated results of the base plate.

$$Nu_3 = 0.54Ra^{0.25} \quad \left(10^4 < Ra < 10^7\right)$$

(15)

From these results, it is confirmed that Equation (15) applied to a range of the low-Rayleigh number. Using Equation (11), the

Nusselt number of the heat sinks was calculated. Here, L was estimated to be the total value of the pin height h_p and the base height 2 mm. As seen this figure, regardless of the pin size, the pin height, the number of pins, it is revealed that Nu and Ra of the heat sink are expressed by an approximate equation, such as Equation (16).

$$Nu = 0.16Ra^{0.52}$$

(16)

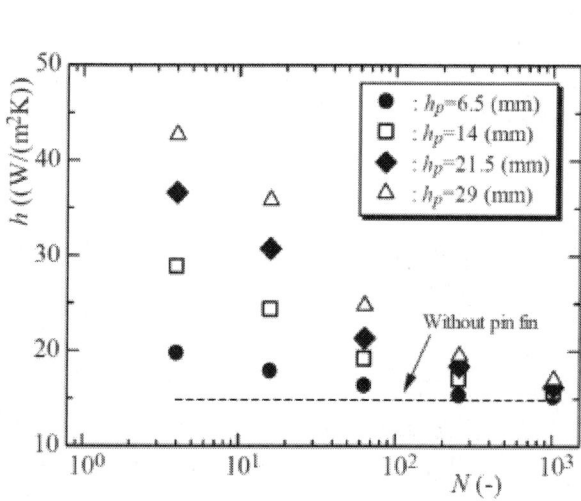

Figure 12: Effects of N and h_p on h (calculated results).

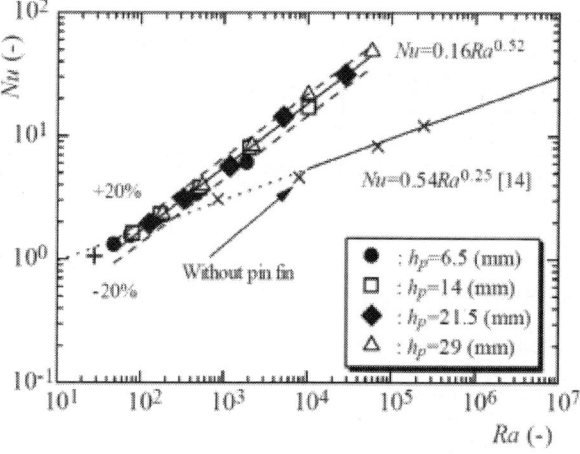

Figure 13: Correlation of Ra and Nu of the heat sink based on the base plate.

Since the calculated results were shown between ±20% of the approximate expression and with miniaturization of pins the Nusselt number Nu of the heat sink comes close to that of the base plate level, it is considered that the proposed correlation implies high accuracy.

CONCLUSIONS

In this study, experimental and numerical studies have been done on the effects of the pin size, pin height, the number of pins on the heat transfer performance of the heat sink.

Concerning the heat sinks with the same heat transfer area, it has been confirmed that the heat sink performance changes depending on the population density of pin and the pin size. From the measured and calculated results, the heat sink temperature has been shown to rise with increase in the number of pins. Especially, the heat sink with miniaturized pins has almost no effect on the heat transfer enhancement. It is considered that these characteristics of the heat sink cause the decrease in the convective component, which occupies approximately 60% of total heat transfer rate, and the heat transfer coefficient due to the decrease in the flow velocity when the pin to pin clearances are reduced.

Regardless of the pin size, the pin height, and the number of pins, it has been revealed that the Nusselt number and the Reyleigh number of the heat sink are represented by an approximate equation.

REFERENCES

1. Aihara, T., Maruyama, S. and Kobayakawa, S. (1990) Free Convective/Radiative Heat Transfer from Pin Fin Arrays with a Vertical Base Plate (General Representation of Heat Transfer Performance). International Journal of Heat Mass Transfer, 33, 1223-1232.http://dx.doi.org/10.1016/0017-9310(90)90253-Q

2. Sara, O.N. (2003) Performance Analysis of Rectangular Ducts with Staggered Square Pin Fins. Energy Conversion and Management, 44, 1787-1803.http://dx.doi.org/10.1016/S0196-8904(02)00185-1

3. Zografos, A.I. and Sunderland, J.E. (1990) Natural Convection from Pin Fin Arrays. Experimental Thermal and Fluid Science, 3, 440-449. http://dx.doi.org/10.1016/0894-1777(90)90042-6

4. Huang, R.T., Sheu, W.J. and Wang, C.C. (2008) Orientation Effect on Natural Convective Performance of Square Pin Fin Heat Sinks. International Journal of Heat and Mass Transfer, 51, 2368-2376. http://dx.doi.org/10.1016/j.ijheatmasstransfer.2007.08.014

5. Sparrow, E.M. and Vemuri, S.B. (1986) Orientation Effects on Natural Convection/Radiation Heat Transfer from Pin- Fin Arrays. International Journal of Heat and Mass Transfer, 29, 359-368. http://dx.doi.org/10.1016/0017-9310(86)90206-1

6. Sertkaya, A.A., Bilir, S. and Kargici, S. (2011) Experimental Investigation of the Effects of Orientation Angle on Heat Transfer Performance of Pin-Finned Surfaces in Natural Convection. Energy, 36, 1513-1517. http://dx.doi.org/10.1016/j.energy.2011.01.014

7. Yu, E. and Joshi, Y. (2002) Heat Transfer Enhancement from Enclosed Discrete Components Using Pin-Fin Heat Sinks. International Journal of Heat and Mass Transfer, 45, 4957-4966. http://dx.doi.org/10.1016/S0017-9310(02)00200-4

8. Bocu, Z. and Altac, Z. (2011) Laminar Natural Convection Heat Transfer and Air Flow in Three-Dimensional Rectangular Enclosures with Pin Arrays Attached to Hot Wall. Applied Thermal Engineering, 31, 3189-3195.http://dx.doi.org/10.1016/j.applthermaleng.2011.05.045

9. Narasimhan, S. and Majdalani, J. (2002) Characterization of Compact Heat Sink Models in Natural Convection. IEEE Transactions on Components and Packaging Technologies, 25, 78-87. http://dx.doi.org/10.1109/6144.991179

10. Minakami, K., Mochisuki, S., Murata, A., Yagi, Y. and Iwasaki, H. (1993) Heat Transfer Characteristics of Pin-Fins with In-Line Arrangement (1st Report, Effect of the Pin Pitch). Japan Society of Mechanical Engineers, 59, 300-307.

11. Kunugi, T., Muko, K. and Shibahara, M. (2004) Ultrahigh Heat Transfer Enhancement Using Nano-Porous Layer. Superlattices and Microstructures, 35, 531-542.http://dx.doi.org/10.1016/j.spmi.2004.04.002

12. Kreith, F. and Bohn, M.S. (2000) Principles of Heat Transfer. 6th Edition, Brooks/Cole.

13. Nakayama, A. (2008) Problems in Heat Transfer. Japan Society of Mechanical Engineers. Maruzen Publishing Co., Ltd.

14. Goldstein, R.J. and Sparrow, E.M. (1973) Natural Convection Mass Transfer Adjacent to Horizontal Plates. International Journal of Heat and Mass Transfer, 16, 1025-1035.http://dx.doi.org/10.1016/0017-9310(73)90041-0

NOMENCLATURES

A: heat transfer area (m²);

c_p: specific heat (J/(kg×K));

Gr: Grashof number (−);

g: gravitational acceleration (m/s²);

h: heat transfer coefficient (W/(m²×K));

h_p: height of pin (mm);

L: characteristic length (m);

N: number of pins (−);

Nu: Nusselt number (−);

P: perimeter (m);

Pr: Prandle number (−);

p: pressure (Pa);

Q: heat transfer rate (W);

Ra: Rayleigh number (−);

T: temperature (°C or K);

u: velocity component in x-direction (m/s);

V: air velocity (m/s);

v: velocity component in y-direction (m/s);

w: velocity component in z-direction (m/s);

w_p: width of pin (mm).

GREEK SYMBOLS

a: thermal diffusivity (m²/s);

b: coefficient of volume expansion (K⁻¹);

e: emissivity (−);

ΔT: temperature difference (°C or K);

λ: thermal conductivity (W/(m×K));

v: kinematic viscosity (m²/s);

ρ: density (kg/m³).

SUBSCRIPTS

a: ambient or air;

c: convection;

g: gauge;

h: representative;

ht: heater;

hs: heat sink;

in: input;

out: output;

r: radiation;

1: lower surface of heated plate;

2: vertical surface of heated plate;

3: upper surface of heated plate.

CITATION

Matsumoto, N. , Tomimura, T. and Koito, Y. (2014) Heat Transfer Characteristics of Square Micro Pin Fins under Natural Convection. Journal of Electronics Cooling and Thermal Control, 4, 59-69. doi: 10.4236/jectc.2014.43007.

Index